Nahomi
Curvelo
336 491 5804

to accompany

Understanding Our Universe

SECOND EDITION

to accompany

Understanding Our Universe

SECOND EDITION

Stacy Palen, Laura Kay, Brad Smith, and George Blumenthal

Steven Desch

GUILFORD TECHNICAL COMMUNITY COLLEGE

 W · W · NORTON & COMPANY · NEW YORK · LONDON

W. W. Norton & Company has been independent since its founding in 1923, when William Warder Norton and Mary D. Herter Norton first published lectures delivered at the People's Institute, the adult education division of New York City's Cooper Union. The Nortons soon expanded their program beyond the Institute, publishing books by celebrated academics from America and abroad. By mid-century, the two major pillars of Norton's publishing program—trade books and college texts—were firmly established. In the 1950s, the Norton family transferred control of the company to its employees, and today—with a staff of four hundred and a comparable number of trade, college, and professional titles published each year—W. W. Norton & Company stands as the largest and oldest publishing house owned wholly by its employees.

Copyright © 2015, 2013, 2010, 2007 by W. W. Norton & Company, Inc.

Printed in the United States of America

Associate Editor, Digital Media: Julia Sammaritano
Project Editor: Diane Cipollone
Director of Production: Jane Searle
Composition by Westchester Publishing Services
Manufacturing by Sheridan Printing

ISBN: 978-0-393-93849-4 (pbk.)

W. W. Norton & Company, Inc., 500 Fifth Avenue, New York, NY 10110
www.wwnorton.com

W. W. Norton & Company, Ltd., Castle House, 75/76 Wells Street, London W1T 3QT
1 2 3 4 5 6 7 8 9 0

CONTENTS

PREFACE

Astronomy is a challenging subject, but a rewarding one. The challenges come in many forms. Because astronomers have learned so much about planets, stars, galaxies, and the universe, there is a lot of material to cover in any college course. Students have many new concepts to learn, and the vocabulary of astrophysics is often strange. Finally, many of the key ideas are hard to visualize because our lives do not give us direct experience of orbits, atomic nuclei, quasars, and the like.

The exercises in this workbook are designed to help you become more familiar with many astronomical phenomena. We begin by thoroughly exploring the appearance of the sky and the motion of the Sun and Moon as seen from Earth. Later, we step out into the Solar System and explore the orbits of the planets and other objects. We then use Starry Night to gather information on the basic properties of stars. Finally, we leave the confines of the Milky Way Galaxy to look out into deep space.

The goal of these exercises is to help you understand many key concepts from the textbook. After completing the assignments, you will be skilled in operating Starry Night and can learn much more on your own.

Please note that the instructions in this manual were written for PC users. The main difference between the controls on a PC and those on a Macintosh is that a right-click on the PC corresponds to CTRL-click on the Macintosh.

A complete guide to the controls can be found by using Starry Night's Help/User's Guide menu. This displays a PDF file. The Quick Tips on page ix of this workbook are meant to summarize the items in the Getting Started and Basics sections of the User's Guide.

QUICK TIPS

- **Orient your gaze toward a compass direction.** Type N, S, E, or W on your keyboard. Pressing the space bar after typing one of these buttons will make the gaze shift instantly to that direction.
- **Change your gaze upward, downward, left, or right.** Use the arrow keys on your keyboard. Alternatively, move the cursor in the display until it looks like a hand; click and drag to change the gaze direction.
- **Zoom in or out toward the center of the display.** Use the + or − buttons, which are located at the lower left of the display.
- **Zoom in or out toward an object on the sky.** Move the cursor toward the object, then use the mouse wheel to zoom in or out.
- **Stop the flow of time.** Click the STOP TIME button (■) in the upper right.
- **Start the flow of time.** Click the RUN TIME FORWARD button (▶). The time step between frames is shown to the left of the time flow buttons and defaults to 1 × (real time).
- **Run the flow of time backward.** Click the RUN TIME BACKWARD button (◀).
- **Advance the time forward by one time step.** Click the STEP TIME FORWARD button (▶|).
- **Move the time backward by one time step.** Click the STEP TIME BACKWARD button (|◀).
- **Multiply the time flow rate.** Click on the three lines to the left of the TIME FLOW RATE display. Select the multiplier from the top part of the drop-down window. For example, to advance at 300 times the normal rate, select "300×."
- **Set a specific time flow rate.** First, select the units for the time flow rate by clicking on three lines to the left of the time step display. Select the desired units

from the drop-down menu. Then click on the number in the TIME FLOW RATE display and type the rate you want. For example, to set the rate to 5 minutes, first select "minutes" from the drop-down menu, then click on the current time step and type **5**.
- **Set a specific date or time.** Move your cursor to the upper left where the time and date are displayed and click on the item you want to change. Either type in the new value or use your keyboard up and down arrows to select one. The right and left arrows are used to move to the next or previous item in the TIME AND DATE display. To change the year from ACE to BCE, click on this item.
- **Find any object.** Type the name of any object in the search bar at the upper right to find information about it or its location in the sky.
- **Open/close a sidebar for more options.** A sidebar can be displayed on the right of the screen to make selecting some display options easier. Click on the menu button (☰) to the left of the search bar to see a menu of useful sidebars. In these exercises, the OPTIONS sidebar will be most commonly used.
- **Expanding menu options.** Within the OPTIONS menu, you will expand the different sections by clicking on the right arrow (▶). When expanded, a section will show a down arrow (▼). The options within a section are turned on/off by use of a toggle button, with the option off when the toggle is to the left (◗) and on when the toggle is to the right (◖).
- **Identify a particular object.** Moving the cursor over an object will bring up a display of information about it. Left-click when pointing at an object, and a label will appear on the display with the name of the object. Right-click to show a menu that will provide other options and information about the object.

to accompany

Understanding Our Universe

SECOND EDITION

The Celestial Sphere

GOAL

- To investigate the apparent daily motion of the sky caused by Earth's rotation.

READING

- Section 2.1 – Earth Spins on Its Axis

We will see what the daily motion of the sky looks like from Earth's North and South poles. This exercise will animate the information shown in Figure 2.5 in the textbook.

It is convenient to imagine the sky as a large sphere centered on Earth. This is called the **celestial sphere**. In reality, the stars and planets are all located at different distances from Earth.

Approximately every 24 hours, the sky appears to rotate around two points. These are called the celestial poles. There are two poles, the **north celestial pole** and the **south celestial pole**. At the North Pole, the north celestial pole is at the **zenith**—the direction straight up.

From either pole, an observer will see one half of the celestial sphere. The half visible from the North Pole is called the northern celestial hemisphere. An observer at the South Pole will see the southern celestial hemisphere. At the poles, all visible stars are above the horizon continually; they neither rise nor set. Such stars are called **circumpolar**.

To navigate the sky, we need to define directions. From any point on the sky, north is the direction from that point toward the north celestial pole. West is the direction of the apparent motion of the celestial sphere (that is, the stars are constantly moving toward the west). East is the opposite direction from west.

Astronomers divide the sky into constellations, which are areas on the sky around particular groups of stars. The stars in each constellation, in general, are not physically associated but instead simply happen to lie in the same direction. Constellations are further discussed in Section 2.2 of the textbook; here, we use them to help visualize the daily motion of the sky.

SETUP

- Start Starry Night.
- Stop the flow of time with the STOP TIME button (■).
- Using the drop-down menus, select OPTIONS / VIEWING LOCATION. A box labeled VIEWING LOCATION will appear.
  - Select the LATITUDE / LONGITUDE tab.
  - In the upper dialog box (LATITUDE), erase the current coordinate and type **90 N**.
  - Click the VIEW FROM SELECTED LOCATION button at the lower right of the dialog box.
  - Your viewing location will now shift to the North Pole. You can speed up the process by hitting the space bar while the motion is taking place. At this point, you may see the sky or you may be looking at the ground.
  - Type the **S** key to face the southern horizon.
- Move the cursor over the window until it becomes a hand. If the horizon does not appear toward the bottom of the window, use the up-arrow key on your keyboard until the horizon appears toward the bottom of the window. Alternatively, you can use the handgrab tool to alter the direction of your gaze;

click and drag upward until the horizon appears at the bottom of the window.

- Click the SIDEBAR button (▯) at the upper right of the window by the search bar to open the sidebar on the right of your window. Then click the menu button (☰) and select OPTIONS to bring up the DISPLAY OPTIONS sidebar.
- If it is not already done, click the right arrow (▶) next to DISPLAY OPTIONS to expand the menu offerings. Set up the following items by expanding each section as needed and clicking the appropriate toggles:
  - GUIDES: CELESTIAL GUIDES: LABELS on, POLES on, other options off.
  - LOCAL VIEW: DAYLIGHT off, LOCAL HORIZON on, other options off.
  - SOLAR SYSTEM: All options off.
  - STARS: STARS on, all other options off.
  - CONSTELLATIONS: LABELS on, BOUNDARIES on, STICK FIGURES (ASTRONOMICAL) on, other options off.
- Close the DISPLAY OPTIONS menu by clicking the SIDEBAR button (▯).

ACTIVITY 1 – DIRECTIONS ON THE SKY

- Set the TIME FLOW RATE to 2 minutes. To do this, click on the menu button (☰) to the left of the TIME FLOW RATE box and select "minutes" from the drop-down menu. Then click on the number in the TIME FLOW RATE box and type **2**.
- Click the ▶ button to start the flow of time. Verify that the time displayed in the upper left is advancing. Watch the motion of the stars as you let time run for a while, then stop the flow of time with the ■ button.

1. At the location of any star, *west* is defined as the direction of the apparent motion, and *east* is the opposite direction. Is west toward the left or toward the right? What about east?
2. Do the stars seem to move parallel to the horizon or at a large angle to the horizon?

ACTIVITY 2 – DIRECTION OF ROTATION

- Move the cursor over the main window and click once. Using the up-arrow key on your computer's keyboard or the handgrab tool, raise the angle of your gaze upward until the north celestial pole is in the middle of the window. You are now looking toward the zenith, and the horizon should be out of view.
- Click on the ▶ button to start the flow of time. Let it run for a while, then stop the action.

3. Is the apparent rotation of the sky clockwise or counterclockwise?
4. In which constellation is the north celestial pole located?

ACTIVITY 3 – VIEW FROM THE SOUTH POLE

- Following the procedures described earlier in the Setup section, set the viewing location to the South Pole, which is at latitude **90 S**. Point your gaze to the north by typing **N**. Use the up arrow on the keyboard or the handgrab tool to place the south celestial pole near the middle of the window. Start the flow of time.

5. Is the apparent rotation of the sky from the South Pole clockwise or counterclockwise?
6. In which constellation is the south celestial pole located?
7. Do stars appear to move in the same direction at the North and South poles? Explain why or why not.

Earth's Rotation Period

GOALS

- To investigate the apparent motion of the sky from an intermediate latitude between the equator and the pole.
- To determine the length of the day as measured by the stars and the Sun.

READING

- Section 2.1 – Earth Spins on Its Axis

This exercise will animate the information shown in Figure 2.8 in the textbook. We will view the motion of the celestial sphere as seen from a place between the equator and the North Pole or the South Pole. From these locations, certain stars will rise and set, while others will remain above the horizon at all times. The latter are termed **circumpolar**.

We also will determine Earth's rotation period, which is the amount of time it takes for Earth to spin once on its axis compared to the distant stars. We do this by noting the times that a particular star crosses the **meridian**. The meridian is a line that runs between the north and south points on the horizon through the **zenith** (the point straight up). The meridian divides the sky into east and west.

Earth's rotation period is called the **sidereal** day. The word *sidereal* refers to the stars. In many other exercises, you will be asked to observe the sky at intervals of one sidereal day or many sidereal days. If you do this, the stars will be in the same position at each observation.

We keep track of time using the Sun, which (on average) crosses the meridian every 24 hours. Intervals of 24 hours are called a **solar day** (or simply a day). In this exercise, we will see that the sidereal day is not 24 hours! The sidereal and solar days differ because Earth orbits the Sun. From Earth, it looks like the Sun moves eastward a little bit compared to the stars each day. The Sun's annual path on the celestial sphere is called the **ecliptic**. The motion of the Sun along the ecliptic is explored in another exercise.

SETUP

- Start Starry Night.
- Stop the flow of time with the STOP TIME button (■).
- If your home location has already been set, then there is no need to change it. Skip to the next dark-bulleted point. If your home location has not previously been set, then follow these instructions to go there.
  - Using the drop-down menus, select OPTIONS / VIEWING LOCATION. A box labeled VIEWING LOCATION will appear.
  - You can select your location from a list of places. To do this, select the LIST tab. Scroll up and down to select a location or type in the name of the nearest major town or city to see if it is included in the list. Pick a location closest to where you live. Click on that location to select.
  - Alternatively, select the MAP tab. Click on a location on the map of Earth near where you live.
  - Click the VIEW FROM SELECTED LOCATION button toward the lower right of the VIEWING LOCATION box.

- Type the **N** key to point your gaze ...ward the north point of the horizon. (Students i... .e Southern Hemisphere should type **S**.)
- Open the OPTIONS sidebar at ... right. Set up the following items by toggling t... .ppropriate options:
  ◦ GUIDES: CELESTIAL GUI... ...: LABELS on, POLES on, other options ...
  ◦ LOCAL VIEW: DAYLIG... ...ff, LOCAL HORIZON on, other options off.
  ◦ SOLAR SYSTEM: All ...ns off.
  ◦ STARS: STARS on, ...er options off.
  ◦ CONSTELLATION... ...BELS, BOUNDARIES, STICK FIGURES ... ...her options off.
- Close the OPTIO... ...bar by clicking the SIDEBAR butt...
- Unless you h... ...ry small window or are observing f... ...igh northern or southern latitude, y... ...ld see both the horizon and the celestial ... your view.

ACTIVIT... CIRCUMPOLAR CONSTELLATIONS

- Set ... ...ME FLOW RATE to 2 minutes. To do this, cl... ...the menu button (≡) to the left of the TIME F... ...RATE box and select "minutes" from the ...down menu. Then click on the number in ...IME FLOW RATE box and type **2**.
- ... ...t the flow of time, and let it run for about ...hours (or longer if necessary), watching the ...oving constellations to see which are mostly ...r entirely circumpolar. Because the option ...AYLIGHT is off, you will see the stars even during the daytime. Stop the flow of time.

... Make a list of constellations that are mostly or entirely circumpolar from your viewing location.

ACTIVITY 2 – RISING AND SETTING CONSTELLATIONS

- Type **E** to change your view so you are facing east.
- Start the flow of time, and let it run for a bit, then stop. You should see stars rise on the eastern horizon.
- Pick a constellation that has just risen in the east, meaning one that is near the horizon and completely visible. Use the STEP TIME BACKWARD button (◀|) to move the time back to the point when the first stars of that constellation are just becoming visible. You can pick the bright stars that are connected by the stick figures. Note the time from the TIME AND DATE display at the upper left.
- Then use the STEP TIME FORWARD button (|▶) to advance the time until the last stars of that constellation are becoming visible. Note the time.

2. What constellation did you pick? How long does it take the whole constellation to rise, in hours?

ACTIVITY 3 – EARTH'S ROTATION PERIOD

- Point toward the southern horizon by typing **S**. (Students in the Southern Hemisphere should type **N** instead.)
- Open the OPTIONS sidebar on the right side of the window.
- Under GUIDES, ALT-AZ GUIDES, select MERIDIAN. You will see a hashed line appear running upward from the direction marker on the horizon.
- Close the OPTIONS sidebar by clicking SIDEBAR button (▢).
- Set the time to be noon (12:00:00 P.M.). Do this by clicking on the hours of the displayed time to highlight it and either typing in the hours you want (**12**) or using the up or down arrows to select the time. Do the same for the minutes and seconds. Change the A.M./P.M. by clicking on it.
- Find a bright star that is near the meridian; any star will do. Move the cursor to the star. When information for that star pops up on your screen, click on the star. The name of the star will appear.
- Set the TIME FLOW RATE to 5 seconds.
- Zoom in so that the area near the star and the meridian are magnified. To zoom, place the cursor near the star you have identified and use the mouse wheel to zoom in or out. (If you do not have a mouse wheel, use the + and – keys in the lower left to zoom.) You may need to adjust your view to keep the star and the meridian both visible near the center of the screen.

3. Use RUN TIME FORWARD (▶) or RUN TIME BACKWARD (◀) to place the star near the meridian and then STOP TIME (■). Then use the STEP TIME FORWARD (|▶) or STEP TIME BACKWARD (◀|) buttons to place the star exactly on the meridian. Write down the time when the star is on the meridian in the table provided for question 3 on your answer sheet.

Now advance the time by 24 hours by clicking on the day value in the current date and then typing the up arrow on your keyboard. (For example, you would change the **5** in the date October **5**, 2007, to **6**.) Alternatively, you can click on the day value and type in a date that is one day later than the current one. USE STEP TIME FORWARD (|▶) or STEP TIME BACKWARD (◀|) until the star is on the meridian again. Write down the time and the interval between this meridian crossing and the previous one.

Do this for 5 more days, filling in the rest of the table. (The first couple of entries are shown as an example only. Your time will vary depending upon your viewing location.) After the table is complete, compute the average interval.

Date	Time	Interval
March 17, 2007	12:26:38 P.M.	
March 18, 2007	12:22:43 P.M.	23 h 56 min 5 s
etc.		
	Average interval	

4. What is the average interval of time between successive crossings of the meridian? This time is Earth's rotation period (1 sidereal day).
5. On average, how many minutes earlier do the stars cross the meridian each day?
6. For a star that crosses the meridian at about midnight tonight, what time will it cross tomorrow night? How about a month (30 days) later? What does this imply about the nighttime visibility of a particular star over the course of a year?

Motion of the Sun along the Ecliptic

GOAL

- To investigate the apparent motion of the Sun during the year caused by the orbit of Earth around the Sun.

READING

- Section 2.2 – Revolution around the Sun Leads to Changes during the Year

As Earth orbits the Sun during the year, the Sun will be located in front of various constellations. This is illustrated in Figure 2.9 of the textbook. In popular astrology, it is said that the Sun "enters Gemini" or "enters Virgo." Many calendars will show the dates the Sun enters each constellation. Astrology columns in the newspapers show the full range of astrological dates that the Sun is in each constellation (for example, Gemini: May 21–June 20).

Each day, the Sun moves about 1° eastward along the celestial sphere, following a path called the **ecliptic**. This ecliptic path is caused by the plane of Earth's orbit around the Sun. As Earth orbits the Sun, our view of it relative to the background constellations constantly changes. The constellations along the ecliptic make up the **zodiac**.

In astrology, the ecliptic is divided evenly among 12 constellations: Gemini, Cancer, Leo, Virgo, Libra, and so on. The dates for each sign actually correspond to the positions of the constellations a very long time ago. Since then, the **precession of the equinoxes**, which is explored in another exercise, has changed the dates corresponding to each astrological sign. Furthermore, astronomers have redefined the boundaries of each constellation in a way that does not correspond exactly to the traditional

boundaries. As a result, the astrological dates for the Sun's position along the zodiac do not correspond to the dates in modern astronomy.

SETUP

- Start Starry Night.
- Stop the flow of time with the STOP TIME button (■).
- Set up the following date: 12:00:00 P.M., July 1, 2007.
- Find the Sun by typing "**Sun**" into the SEARCH BAR at the upper right.
  - In the results, the Sun will be at the top of the list. Click the ☑ box to its left to label the Sun in the sky.
  - Click on the menu button (≡) to the left of the word SUN in the results to show a drop-down menu. Click CENTER; this will aim your point of view so that the Sun is centered along the line of sight.
- Open the OPTIONS sidebar using the menu next to the SEARCH BAR.
  - Under GUIDES, have all options off.
  - Under LOCAL VIEW, turn off DAYLIGHT and turn on LOCAL HORIZON. All others should be off.
  - Under SOLAR SYSTEM, turn PLANETS-MOONS on. All other options should be off.
  - Under STARS, turn STARS on and MILKY WAY off.
  - Under CONSTELLATIONS, turn LABELS, BOUNDARIES, and STICK FIGURES (ASTRONOMICAL) on. All other options should be off.
- Close the OPTIONS sidebar.
- Set the TIME FLOW RATE to be 1 day. The units ("days") are selected with the drop-down menu next

to the TIME FLOW RATE, and it will default to 1 when selected.

- At the bottom left, use the + and − buttons to zoom out as necessary. At this point, you should see that the Sun is located within the boundaries of Gemini.
- Now click the STEP TIME BACKWARD button (◀) once. The TIME AND DATE display at the upper left should read noon on the previous day (June 30, 2007). Keep clicking backward until the Sun is well within Taurus. Experiment with going backward and forward 1 day at a time using this button and the STEP TIME FORWARD (▶) button, and then set the Sun to the boundary between Taurus and Gemini.

Constellation	Date Sun Enters
Gemini	May 21
Cancer	June 21
Leo	July 21
Virgo	August 21
Libra	September 21
Scorpius	October 21
Sagittarius	November 21
Capricornus	December 21
Aquarius	January 21
Pisces	February 21
Aries	March 21
Taurus	April 21

ACTIVITY 1 – THE SUN AND THE ZODIAC

1. Use the STEP TIME FORWARD button to advance the time 1 day at a time. In the table on your answer sheet, keep track of the day the Sun enters a new constellation and then the number of days that the Sun spends in that constellation. The first two constellations are done as examples, starting with the Sun entering Gemini.

Constellation	Date Sun Enters	Days Spent
Gemini	6/22/2007	29
Cancer	7/21/2007	21
etc.		

After a year, you will see that the Sun enters Gemini again. Do not just copy down the table of traditional constellations and dates that is given in the next question. There are major differences now!

ACTIVITY 2 – ASTROLOGICAL DATES

2. Compare your table of dates to the traditional astrological dates in the table that follows. What differences do you find between the traditional and the modern astronomical dates?

ACTIVITY 3 – DATES IN THE DISTANT PAST

Reset the date to May 21, 2007. According to the traditional astrological dates, this is when the Sun should enter Gemini; however, it is still well within Taurus. Now enter 1500 ACE for the year. You will notice that the orientation of the constellations has changed as a consequence of precession and that the Sun is still within Taurus but is closer to Gemini.

Using the STEP TIME FORWARD (▶) button, you will see that, in this year, the Sun entered Gemini about June 5, which is closer to the traditional date of May 21.

Reset the date to May 21 and then try even earlier years. Keeping the date set to May 21, find a year (there will be a range of correct years) when the Sun entered Gemini about May 21. (Hint: You may have to go back to BCE for this.) There's no need to be exact—use century years (for example, 100 ACE, 1 ACE, 100 BCE) until you find the right date.

3. In what year did the Sun enter Gemini about May 21?
4. Why does the Sun only appear to pass through the constellations of the zodiac? For example, why doesn't the zodiac include Ursa Major?

Name: _____

Class/Section: _____

Starry Night Student Exercise – Answer Sheet
Motion of the Sun along the Ecliptic

1. Table of dates:

Constellation	Date Sun Enters	Days Spent
Gemini		
Cancer		

2. What differences do you find between the traditional and the modern astronomical dates?

_____

_____

_____

3. In what year did the Sun enter Gemini about May 21?

_____

4. Why does the Sun only appear to pass through the constellations of the zodiac? For example, why doesn't the zodiac include Ursa Major?

_____

_____

Motion of the Moon

GOALS

- To investigate the apparent motion of the Moon during each month.
- To determine the interval between successive moonrises.
- To measure the Moon's orbital period and the time for lunar phases to repeat.

READING

- Section 2.3 – The Moon's Appearance Changes as It Orbits Earth

The Moon completes one orbit around Earth in approximately 1 month. If you were keeping track of the Moon's position in the sky, then you could measure this time by watching how long the Moon takes to return to a particular place in the celestial sphere, such as a fixed star. This length of time, which is 27.32 days, is the Moon's **sidereal** period, the word *sidereal* referring to the stars.

The amount of time it takes the Moon to complete a cycle of phases, such as from new Moon back to new Moon, is called the Moon's **synodic** period. The synodic period of 29.53 days is slightly longer than the Moon's sidereal period because as the Moon orbits Earth, Earth is also orbiting the Sun. In the time it takes the Moon to complete one full orbit, Earth has moved slightly around the Sun, so it will take a little bit longer for the Moon to catch up to the Sun's new location in the celestial sphere.

In this exercise, we will obtain both the sidereal and synodic periods by actual measurements and compare these to the given values. We will also determine the amount of time between moonrises on successive nights.

SETUP

- Start Starry Night.
- Stop the flow of time with the STOP TIME button (■).
- Set up the following date: 3:00:00 A.M., December 12, 2006.
- Make your horizon flat and free of obstructions (if it is not already). Open the OPTIONS drop-down menu at the top left of the screen (between the "View" and "Labels" menus), and choose OTHER OPTIONS / LOCAL HORIZON. In this window, click the button next to FLAT and click OK.
- Aim to the eastern point of the horizon (type the **E** key).
- Open the OPTIONS sidebar on the right.
  - Under GUIDES, have all options off.
  - Under LOCAL VIEW, turn off DAYLIGHT, turn on LOCAL HORIZON, and turn all others off.
  - Under SOLAR SYSTEM, turn PLANETS-MOONS on, turn all other options off.
  - Under STARS, turn STARS on, turn all other options off.
  - Under CONSTELLATIONS, turn LABELS and STICK FIGURES on, all other options off.
- In the SEARCH BAR at the upper right, type "**The Moon**."
  - In the results, the Moon will be at the top of the list. Click the ☑ box to its left to label the Moon in the sky.
- Close the OPTIONS sidebar.
- Set the TIME FLOW RATE to be 2 minutes. Don't start the flow of time yet.

ACTIVITY 1 – TIME OF MOONRISE ON SUCCESSIVE NIGHTS

- Depending on your viewing location, you should see that the Moon is located near the horizon. It may be above the horizon, which means it rose earlier than 3:00 A.M. local time. If it is below the horizon, it will rise later.
- If the Moon is above the horizon, use the STEP TIME BACKWARD button (◀). Keep going until the Moon is located right on the horizon. This will give you the time of moonrise. Write down this time in the table on your answer sheet.
- Alternatively, if the Moon has not yet risen, use the STEP TIME FORWARD (▶) button and record the time of moonrise.
- Advance the time using the RUN TIME FORWARD button (▶). Let 1 day (24 hours) pass; then stop the flow of time with the STOP TIME button (■). Then step the time forward or backward, as appropriate, until the Moon is on the eastern horizon. Record the time of moonrise for the next date in your table.
- Do this for 5 more days (that is, until moonrise on December 17). You should note that the Moon rises later each night.

1. Fill out the table in question 1 of the answer sheet showing the time of moonrise on each night. It will look like this:

Night	Date	Time of Moonrise	Interval between Moonrises
0	Dec. 12	1:57 A.M.	–
1	Dec. 13	2:59 A.M.	24 h 62 min
2	etc.		
3			
4			
5			
		Average interval	

In the last column of the table, fill in the intervals between moonrises. The exact times you get for the moonrises will be different, as they depend on your viewing location. Express the intervals as 24 hours and some number of minutes to make finding the average easier.

2. Each interval should be greater than 24 hours, meaning that the Moon rises later each night. What is the average number of minutes beyond 24 hours that the Moon rises later each night?

ACTIVITY 2 – THE MOON'S SIDEREAL PERIOD

- Reset the date to December 12, 2006, and the time to 3:00 A.M.

- Type **S** to change your view to the southern horizon. (Students in the Southern Hemisphere should type **N**.)
- Press the RUN TIME FORWARD button (▶) and let time pass until the Moon is above the southern horizon, about halfway between the bottom and the top of the window, and then stop.
- Point at a star near the Moon with the cursor. When its information pops up on the screen, click on the star. Its name will appear.
- If the label "The Moon" disappears, type this back into the search bar and click the ☑ box to the left of the Moon to make it come back.
- Write down the initial date and the time when the Moon and this star are close to each other in the sky.
- Set the TIME FLOW RATE to be 1 sidereal day. Click once on STEP TIME FORWARD (▶). You will note that the Moon has moved eastward against the stars, which stay in the same place in the sky. Keep clicking until the Moon returns closest to the star you marked. Write down the date and the time.

3. According to your observations, how long does it take for the Moon to return to the same position compared to the stars? (If the closest approach is midway between two dates, estimate the result.)

4. What is the value of the Moon's sidereal period given in the Reading section of this exercise? How accurate was your answer in the previous problem?

ACTIVITY 3 – THE MOON'S SYNODIC PERIOD

- Open the OPTIONS sidebar.
  - Under GUIDES, ALT-AZ GUIDES, select MERIDIAN.
  - Under LOCAL VIEW, turn on DAYLIGHT.
- Find the Sun by typing "**Sun**" into the SEARCH BAR at the upper right.
  - In the results, the Sun will be at the top of the list. Click the ☑ box to its left to label the Sun in the sky.
- Set TIME FLOW RATE to 2 minutes.
- Advance the time until the Sun is located at the local meridian.
- Set the TIME FLOW RATE to 1 day.
- Click STEP TIME FORWARD (▶) or STEP TIME BACKWARD (◀) until the Moon is near the Sun. Write down this date. At this time, there is a new Moon because the Sun and the Moon are aligned in the sky.
- Click forward 1 day. Note that the Moon moves to the east away from the Sun. Keep track of how many days you must advance time until the Moon

is closest to the Sun again, about a month later. Write down this date.

5. According to your observations, how long does it take the Moon to return to the same position compared to the Sun?

6. What is the value of the Moon's synodic period given in the Reading section of this exercise? How accurate was your answer in the previous problem?

7. What is the cause of the difference between the Moon's sidereal period and its synodic period? Why is one longer than the other?

4:06 am
January 8   2:19 am

January 18, 2007

STUDENT EXERCISE | Precession

GOAL

- To demonstrate the precession of Earth's axis.

READING

- Section 2.2, last section – Earth's Axis Wobbles

We look more closely at the location of the north celestial pole (NCP) with respect to the stars and see that it slowly changes over many years. This is due to the **precession** of Earth's axis. In the textbook, the precession is described in terms of the changing location of the equinoxes, but in this exercise we will focus on the pole. The key concepts for precession are illustrated in Figure 2.12 of the text book.

Recall that the NCP is not an object in space, but rather a direction. An observer at Earth's North Pole will see that the NCP is at the zenith (that is, straight up). An observer at another latitude will see the celestial pole at an angle above the northern horizon (for example, Figure 2.8).

Earth's rotation axis wobbles slowly. Over a period of about 26,000 years, it traces out a circle on the sky that is about 47° in diameter. In this exercise, we watch the changing position of the NCP over a few thousand years.

SETUP

- Start Starry Night.
- Stop the flow of time with the STOP TIME button (■).
- Using the drop-down menus, select OPTIONS / VIEWING LOCATION. A box labeled VIEWING LOCATION will appear.

- Select the LATITUDE / LONGITUDE tab.
- In the upper dialog box (LATITUDE), remove the current coordinate and type **55 N**.
- Click the VIEW FROM SELECTED LOCATION button at the lower right of the dialog box.
- Your viewing location will now shift. You can speed up the process by hitting the space bar while the motion is taking place. At this point, you may see the sky or you may be looking at the ground.
- Type **N** to point your gaze toward the northern horizon.
- Open the OPTIONS sidebar.
  - Under GUIDES / CELESTIAL GUIDES, turn POLES on.
  - Under LOCAL VIEW, turn on LOCAL HORIZON; all other options should be off.
  - Under SOLAR SYSTEM, all options should be off.
  - Under STARS, turn STARS on; all other options should be off.
  - Under CONSTELLATIONS, turn on LABELS, BOUNDARIES, and STICK FIGURES; all other options should be off.
- Close the OPTIONS sidebar.

ACTIVITY 1 – THE DISTANCE BETWEEN POLARIS AND THE NORTH CELESTIAL POLE

- Identify Polaris and click on it with the left mouse button to display its name. (Polaris is the bright star at the end of the handle of the Little Dipper in the constellation Ursa Minor, right next to the north celestial pole; use the stick figures for the constellation as a guide.) *Do not* use the CENTER option (found on the right-click drop-down menu) to center on Polaris.

- Zoom in toward the NCP using the + or – buttons (bottom left) or the cursor wheel. You may have to grab and drag the screen to move the NCP so that it is near the center of the window. Zoom in so that the NCP and Polaris are well separated.
- Use the angle measuring tool to find the separation between the NCP and Polaris. To do this, click directly on the dot that represents Polaris and drag the mouse toward the NCP. If you did not point directly at the star, then the screen will be dragged instead. When the cursor is clicked right on the star, a red line and an angular separation measurement will appear on the screen that can be dragged around to measure separations. Drag the measuring tool to the little cross representing the NCP, and read off the separation. The angular separation is listed in degrees, minutes, and seconds (1 minute is 1/60 of a degree; 1 second is 1/60 of a minute, or 1/3,600 of a degree.)

1. What value do you get for the current separation between Polaris and the NCP?

ACTIVITY 2 – WHEN POLARIS IS CLOSEST TO THE NCP

- Change the time flow rate to 366 sidereal days.
- Click RUN TIME FORWARD (▶). You will see that the position of the NCP relative to Polaris slowly changes.
- When you estimate that Polaris is closest to the NCP, click STOP TIME (■). You may wish to go backward and forward until you decide that the two are closest together.

2. Approximately when (within 50 years) will Polaris and the NCP be closest together?

- Measure the angular separation between the NCP and Polaris when they are at their closest. Divide this measurement by your answer for the current separation from question 1, and multiply by 100 to convert this to a percentage. (Use only the minute values in this calculation.)

3. What will the separation be compared to now, in percent?

ACTIVITY 3 – THE SEPARATION IN THE PAST

- Long before compasses were invented, navigators in the Northern Hemisphere used Polaris as the "North Star" to orient themselves. Set the TIME AND DATE values so that they show some date in the year 1 ACE.

4. What was the separation between Polaris and the NCP in the year 1 ACE?

- Consider the stars connected by the stick figures in the constellations of Ursa Minor and Draco. Try to identify stars in these two constellations that were closer to the NCP than Polaris was, and so would make a better North Star in this year. Select a candidate star by moving the cursor over it to display its name on the screen. Then use the angular separation tool to find the separation between this star and the NCP.

5. List three stars that were closer than Polaris to the NCP in the year 1 ACE. What was the angular separation between each star and the NCP?
6. Why was Polaris still used as the "North Star" even when there were other stars that were closer?
7. Does the change in the position of the north celestial pole have any effect on the other stars in the sky besides Polaris? Why or why not?

Name: _____

Class/Section: _____

Starry Night Student Exercise – Answer Sheet
Precession

1. What value do you get for the current separation between Polaris and the NCP?

   _____

2. Approximately when (within 50 years) will Polaris and the NCP be closest together?

   _____

3. What will the separation be compared to now, in percent?

   _____

4. What was the separation between Polaris and the NCP in the year AD 1?

   _____

5. List three stars that were closer than Polaris to the NCP in the year 1 ACE. What was the angular separation between each star and the NCP?

   _____

   _____

   _____

6. Why was Polaris still used as the "North Star" even when there were other stars that were closer?

   _____

7. Does the change in the position of the north celestial pole have any effect on the other stars in the sky besides Polaris? Why or why not?

   _____

   _____

STUDENT EXERCISE | Kepler's Laws

GOAL

- To demonstrate Kepler's second and third laws for planetary orbits.

READING

- Section 3.1 – Since Ancient Times Astronomers Have Studied the Motions of the Planets

Based on accurate data from Tycho Brahe, Kepler advanced three empirical **laws** describing the orbits of the planets. In particular, he came up with rules relating the size of an orbit to the orbital period (the third law) and described the speed a body follows along its orbital path (the second law). The exact wording of these laws can be found in Section 3.1 of the textbook. The word **empirical** means that the laws are derived from observations and not from physical theory.

Each planet orbits the Sun along a path that is in the shape of an **ellipse**. The size of an orbit is given by the length of the **semimajor axis**. This is half the longest dimension of the orbit. Each planet will have a minimum and a maximum distance from the Sun called, respectively, the **perihelion** and the **aphelion**.

Most of the major planets have orbits that are almost circular; they travel at nearly a constant speed along the orbit. In this exercise, we will create a new asteroid and give it a very elongated orbit, which will make its changing speed more apparent.

SETUP

- Start Starry Night.
- Stop the flow of time with the STOP TIME button (■).
- Open the OPTIONS sidebar.
  - Under LOCAL VIEW, turn all options off.
  - Under SOLAR SYSTEM, select PLANETS-MOONS and ASTEROIDS, and turn all other options off.
  - Under STARS, turn all options off.
  - Under CONSTELLATIONS, turn all options off.
  - Under DEEP SPACE, turn all options off.
- Type "**Sun**" in the SEARCH BAR at the upper right.
  - Double-click on the word SUN. This will shift your gaze so you are pointed at the Sun and will put a label near the Sun.
  - Earth should also be in the search results. If not, type "**Earth**" in the SEARCH BAR. Click the box to the left of Earth to label it and the circle to the far right of Earth to show its orbit.
- Select OPTIONS / VIEWING LOCATION from the drop-down menus at the top of the Starry Night window. Click on the down arrow (▼) next to the display box for VIEW FROM. Select STATIONARY LOCATION.
- Under CARTESIAN COORDINATES, replace the current numbers with
  - X: 0 AU
  - Y: 0 AU
  - Z: 4 AU

then click the VIEW FROM SELECTED LOCATION button. The view will now shift so that you are looking down on the Sun and Earth's orbit from a position 4 AU above the Sun.

ACTIVITY 1 – KEPLER'S THIRD LAW

- We will create a new asteroid that has an elongated orbit. Using the drop-down menus at the top of the Starry Night window, select FILE / NEW / NEW ASTEROID ORBITING SUN. A subwindow labeled ASTEROID: UNTITLED will appear.
  - There will be a dialog box at the top that has the word UNTITLED in it. Replace what is there with a capital **X**. This is the name of the new asteroid. You can call it something else, as long as it is not the same as one of the asteroids already on the list.
  - The subwindow will show a series of slider bars under a tab called ORBITAL ELEMENTS. To the right of each slider bar is a dialog box in which you can enter numbers manually. Enter the following numbers:
    Mean distance (a): 1.0
    Eccentricity (e): 0.75
  - Leave the other numbers alone. Close the subwindow by clicking on the × at the top right corner. You will be asked whether you want to save the new information. Select SAVE.
- If the sidebar at right has been closed, open it again by typing "**Sun**" in the SEARCH BAR. Scroll all the way to the bottom to find a section labeled USER CREATED list. The name of the asteroid you created (X) will be the only entry in this list.
- Click the box to the left of X to label it and the circle to the far right of X to show its orbit.

1. The value of a you entered is the length of the semimajor axis of X's orbit. This asteroid has the same value of a as does Earth. Use Kepler's third law, which can be found in Section 3.1 of the textbook, to determine its orbital period. What is the orbital period of asteroid X?
2. Describe how the period of the orbit depends on the orbital eccentricity, e.
3. If we had created an orbit with $a = 4$ AU, what would the period be?

ACTIVITY 2 – THE PERIOD OF X'S ORBIT

- Earth's orbit is the green circle displayed on screen. Zoom in using +/– to make the orbit fill the Starry Night display. You will notice that there are two places where X's orbit crosses Earth's.
- Set the TIME FLOW RATE to 1 day. Using the motion buttons, advance or reverse the time until X crosses Earth's orbit. Stop the flow of time.

- Use the STEP TIME FORWARD (▶|) or STEP TIME BACKWARD (|◀) buttons until you find the exact date on which X was closest to Earth's orbit.
- Write down the date from the TIME AND DATE display.
- Now click RUN TIME FORWARD (▶) and let X orbit the Sun completely. Stop the time when it has gone all the way around its orbit and is crossing Earth's orbit again. You have to pick the original crossing point and ignore the other one. Line up X with Earth's orbit as you did in the previous step. Write down the date.

4. How long did it take for X to complete an orbit? Does your answer agree with your computation in question 1?
5. Is X moving faster at perihelion or at aphelion?
6. State Kepler's second law as applied to a planet's orbital speed. Does your answer for question 5 agree with this law?

ACTIVITY 3 – INSIDE AND OUTSIDE

- Now determine the date on which X crosses Earth's orbit going inward and the date when it crosses Earth's orbit going outward.

7. How many months does X spend inside Earth's orbit? How many months does it spend outside Earth's orbit?

ACTIVITY 4 – KEPLER'S SECOND LAW

- Click the menu button (≡) to the left of X under the USER CREATED column and select SHOW INFO.
- Under POSITION IN SPACE, find the display for DISTANCE FROM THE SUN. Use the time flow buttons to move X until it is at perihelion. Write down the distance. Do the same for aphelion.

8. What is the distance from the Sun at perihelion and at aphelion?
9. If the semimajor axis for X is 1 AU, what is the length of the major axis? Does this agree with the numbers you measured in question 8?
10. If you wanted to find all of the potential hazardous asteroids which have orbits that cross Earth's orbit, such as asteroid X, where in the Solar System are you most likely to find them at any given time? Thus, how would you focus your observational search for such objects? (Hint: Think of what Kepler's second law says about elliptical orbits.)

Name: _____

Class/Section: _____

Starry Night Student Exercise – Answer Sheet
Kepler's Laws

1. What is the orbital period of asteroid X as calculated from Kepler's law?

   _____

2. Describe how the period of the orbit depends on the orbital eccentricity, *e*.

   _____

3. If we had created an orbit with $a = 4$ AU, what would the period be?

   _____

4. How long did it take for X to complete an orbit? Does your answer agree with your computation in question 1?

   _____

5. Is X moving faster at perihelion or at aphelion?

   _____

6. State Kepler's second law as applied to a planet's orbital speed. Does your answer for question 5 agree with this law?

   _____

7. How many months does X spend inside Earth's orbit? How many months does it spend outside Earth's orbit?

   _____

8. What is the distance from the Sun at perihelion and at aphelion?

   _____

9. If the semimajor axis for X is 1 AU, what is the length of the major axis? Does this agree with the numbers you measured in question 8?

   _____

10. If you wanted to find all of the potential hazardous asteroids which have orbits that cross Earth's orbit, such as asteroid X, where in the Solar System are you most likely to find them at any given time? Thus, how would you focus your observational search for such objects? (Hint: Think of what Kepler's second law says about elliptical orbits.)

    _____

    _____

Flying to Mars

GOAL

- To create an orbit that takes a spacecraft from Earth to Mars.

READING

- Section 3.1 – Since Ancient Times Astronomers Have Studied the Motions of the Planets

Kepler's laws of planetary motion have many applications besides explaining the orbits of the planets, asteroids, and comets. In this exercise, we will see how we can set up an orbit that will carry a spacecraft from Earth to the other planets. We will do this by creating an asteroid in the Starry Night database and adjusting the parameters of its orbit until it connects Earth's orbit with that of Mars. Such an orbit is called a Hohmann transfer orbit, and generally it minimizes the amount of fuel needed to reach another planet. We will also find that spacecraft have to be launched at particular times in order to reach their destinations.

We will also consider an important complication. If the spacecraft has a human crew aboard, the timing gets very tricky indeed! Not only does the spacecraft have to be launched at just the right time to reach Mars, but also Earth might not be in the right place when the crew wants to return.

SETUP

- Start Starry Night.
- Stop the flow of time with the STOP TIME button (■).

- Open the OPTIONS sidebar.
  - Under GUIDES, turn all options off.
  - Under LOCAL VIEW, turn all options off.
  - Under SOLAR SYSTEM, select PLANETS-MOONS and ASTEROIDS and turn all other options off.
  - Under STARS, turn all options off.
  - Under CONSTELLATIONS, turn all options off.
  - Under DEEP SPACE, turn all options off.
- Type "**Sun**" in the SEARCH BAR at the upper right.
  - Double-click on the word SUN. This will shift your gaze so you are pointed at the Sun and will put a label near the Sun.
  - Earth should also be in the search results along with other Solar System objects. If not, type "**Earth**" in the SEARCH BAR. Click the box to the left of Earth to label it and the circle to the far right of Earth to show its orbit.
- Repeat the same procedure to find Mars if it is not already shown. Click the box to the left of Mars to label it and the circle to the far right of Mars to show its orbit.
- Select OPTIONS / VIEWING LOCATION from the drop-down menus at the top of the Starry Night window. Click on the down arrow (▼) next to the display box for VIEW FROM. Select STATIONARY LOCATION.
- Under CARTESIAN COORDINATES, replace the current numbers with
      X: 0 AU
      Y: 0 AU
      Z: 4 AU
  then click the VIEW FROM SELECTED LOCATION button. The view will now shift so that you are looking down on the Sun and Earth's orbit from a position 4 AU above the Sun.

- If you completed the exercise "Kepler's Laws," you will have created an asteroid called X. If not, then do the following steps:
- We will create a new asteroid that has an elongated orbit. Using the drop-down menus at the top of the Starry Night window, select FILE / NEW / NEW ASTEROID ORBITING SUN. A subwindow labeled ASTEROID: UNTITLED will appear.
  - There will be a dialog box at the top that has the word UNTITLED in it. Replace what is there with a capital **X**. This is the name of the new asteroid. You can call it something else, as long as it is not the same as one of the asteroids already on the list.
  - The subwindow will show a series of slider bars under a tab called ORBITAL ELEMENTS. To the right of each slider bar is a dialog box in which you can enter numbers manually. Enter the following numbers:
    > Mean distance (a): 1.0
    > Eccentricity (e): 0.75
  - Leave the other numbers alone. Close the subwindow by clicking on the × at the top right corner. You will be asked whether you want to save the new information. Select SAVE.
- If the sidebar at right has been closed, open it again by typing "**Sun**" in the SEARCH BAR. Scroll all the way to the bottom to find a section labeled USER CREATED list. The name of the asteroid you created (X) will be the only entry in this list.
- Click the box to the left of X to label it and the circle to the far right of X to show its orbit.

ACTIVITY 1 – THE TRANSFER ORBIT

- Set the date to read June 28, 2003.
- Using the zoom in/out buttons or your mouse wheel, zoom in until the orbit of Mars fills the display window.
- Find the name X under the list of User Created Asteroids. Click on the menu button (≡) to the left of X and select EDIT ORBITAL ELEMENTS.
- Using the various slider bars, adjust the MEAN DISTANCE (a), the ECCENTRICITY (e), and the ARG[UMENT] OF PERICENTER (w) values until the orbit of X touches Earth's orbit at X's perihelion and Mars's orbit at X's aphelion. In particular, you want the perihelion of X to be at the same place as Earth

was on June 28, 2003. First start with values close to
  - Mean distance (a): 1.25 AU
  - Eccentricity (e): 0.20
  - Arg[ument] of pericenter (w): 270°
Using the slider bars, make small adjustments so that the orbit of X connects Earth's orbit with Mars's orbit.
- Now adjust the value of the MEAN ANOMALY (L) until the asteroid X is located where Earth was on June 28, 2003. It should be near 220°.

1. What value did you get for the eccentricity (e) for X's orbit?
2. What is the mean distance (a)?
3. What is the value of the argument of pericenter (w)?
4. What is the value of the mean anomaly (L)?

ACTIVITY 2 – TIME TO FLY TO MARS

- Click on the × at the upper right corner of the subwindow to close it.
- Set the TIME FLOW RATE to 1 day. Start the flow of time with the ▶ button and let it run. You will see that X follows the orbit you set up previously. You should see that the spacecraft gets close to Mars after several months. Don't worry if it misses a little bit—it's not easy to set up an exact orbit that places the spacecraft exactly at Mars's location. (NASA has fancy computers for that!) Use the STOP TIME (■) and the STEP TIME (◀ or ▶) buttons to settle on a time when X is closest to Mars.

5. Approximately when does the spacecraft arrive in the vicinity of Mars?

ACTIVITY 3 – COMING HOME

- Now continue the flow of time with the ▶ button. The spacecraft will return to Earth's orbit.

6. When does X return to Earth's orbit? Is Earth there at the time?
7. What is the orbital period of X?
8. Suppose X was a spacecraft with a crew aboard. As a mission planner, what might you have to do to make sure the crew gets back to Earth safely?

Name: _____

Class/Section: _____

Starry Night Student Exercise – Answer Sheet
Flying to Mars

1. What value did you get for the eccentricity (*e*) for X's orbit?

   _____

2. What is the mean distance (*a*)?

   _____

3. What is the value of the argument of pericenter (*w*)?

   _____

4. What is the value of the mean anomaly (*L*)?

   _____

5. Approximately when does the spacecraft arrive in the vicinity of Mars?

   _____

6. When does X return to Earth's orbit? Is Earth there at the time?

   _____

7. What is the orbital period of X?

   _____

8. Suppose X was a spacecraft with a crew aboard. As a mission planner, what might you have to do to make sure the crew gets back to Earth safely?

   _____

   _____

   _____

STUDENT EXERCISE | The Moons of Jupiter

GOAL

- To investigate historical observations of Jupiter and its four large moons and to use these observations to deduce the motion of these moons around the planet.

READING

- Section 3.1 – Since Ancient Times Astronomers Have Studied the Motions of the Planets
- Section 9.2 – Moons as Small Worlds

In the year 1610, Galileo first turned a telescope to the planet Jupiter. What he found there were four small "stars" that appeared to move along with Jupiter, constantly changing their positions relative to the planet. Over the course of several weeks in January of that year, he methodically recorded the changing positions of these "stars," sometimes observing them more than once per night. That he was able to record accurately the positions of what we now understand to be four moons is especially amazing considering the rudimentary telescope he used.

By using his observations, Galileo was able to deduce that these moons were in orbit around Jupiter and were not just stars in the background. This was a revolutionary idea at a time when everything was thought to orbit Earth. But this was not a simple deduction to make. In this exercise, we will first re-create Galileo's observations using the dates and times he recorded. Then we will find the moons' orbital periods and deduce the relation between their distances, speeds, and periods.

SETUP

- Start Starry Night.
- Stop the flow of time with the STOP TIME button (■).
- Open the OPTIONS sidebar.
  - Under GUIDES, all options should be off.
  - Under LOCAL VIEW, turn on DAYLIGHT and LOCAL HORIZON. All others should be off.
  - Under SOLAR SYSTEM, turn on PLANETS-MOONS.
  - Under STARS, turn on STARS. All others should be turned off.
  - Expand the STARS options by clicking the down arrow at left.
  - Turn on LIMIT BY MAGNITUDE. Then move the cursor over LIMIT BY MAGNITUDE. The label will turn into LIMIT BY MAGNITUDE OPTIONS. Click on it.
  - A window titled STARS OPTIONS will appear. In this window, click the ☑ box next to LIMIT STARS BY MAGNITUDE.
  - In the box to the right of the slider bar, type in **7.00**.
  - Click OK to exit this window.
  - Under DEEP SPACE, turn all options off.
- Close the OPTIONS sidebar.

ACTIVITY 1 – GALILEO'S OBSERVATIONS

- Using the drop-down menus, select OPTIONS / VIEWING LOCATION. A box labeled VIEWING LOCATION will appear.
  - In the LIST tab, type "**Padova**," which will bring up the location name Padova, Veneto, Italy, the city where Galileo made his observations.

- ○ Click on this location name to select it.
- ○ Click the VIEW FROM SELECTED LOCATION button in the lower right corner of the dialog box.
- Type "**Jupiter**" in the SEARCH BAR.
  - ○ Click on the menu button (☰) to the left of the word JUPITER to show a drop-down menu. Click CENTER; this will aim your point of view so that Jupiter is centered along the line of sight.
- Close the OPTIONS sidebar.
- Set the following date and time: 8:00:00 P.M., January 7, 1610 AD.
- Use the + button at the bottom left to zoom in until the display reads 1° × 1° or as close as you can to it.

1. Your view is now set up to display what Galileo would have seen through his telescope on January 7, 1610. In fact, his view was much more limited because of the small field of view of this telescope. His telescope also did not have a large magnification, which is why Jupiter looks so small, making it a challenge to identify the moons. His first observation as taken from his own records is completed for you in the table. Notice that he only recorded three of the moons. Continue filling in the table using the dates and times of Galileo's next six observations. (Note that January 9 was skipped due to poor weather.)

Date	Time	Sketch		
1/07/1610	8.00 P.M.	*	* ○	*
1/08/1610	8:00 P.M.			
1/10/1610	8:00 P.M.			
1/11/1610	8:00 P.M.			
1/12/1610	8:00 P.M.			
1/13/1610	8:00 P.M.			
1/14/1610	8:00 P.M.			

2. In his early observations, Galileo only saw three moons. It was not until January 13, 1610, that he recorded the fourth. Looking at your observations, can you propose an explanation for why Galileo may not have immediately noticed all four moons? Use one of your own observations to support your answer.

ACTIVITY 2 – THE ORBITS OF JUPITER'S MOONS

- Using the drop-down menus, select OPTIONS / VIEWING LOCATION. A box labeled VIEWING LOCATION will appear.
  - ○ Click on the down arrow (▼) next to the display box for VIEW FROM. Select STATIONARY LOCATION.

- Under CARTESIAN COORDINATES, replace the current numbers with
  - X: 1 AU
  - Y: 0 AU
  - Z: 0 AU

  Then click the GO TO LOCATION button. The view will now shift so that you are looking out into space at a position 1 AU from the Sun.
- Click the ▢ button to open the sidebar again. Your search results for Jupiter will still be there.
- Click the menu button (☰) to the left of Jupiter to show a drop-down menu. Click CENTER; this will aim your point of view so that Jupiter is in the center of your screen again.
- The four Galilean moons (Io, Europa, Ganymede, and Callisto) should already be listed below Jupiter. If not, click the down arrow (▼) to expand Jupiter's list.
- Click the checkbox to the left of each of these moons to label them on the display.
- Close the sidebar.
- At the top right, use the + button to zoom in until Jupiter and all four of the Galilean moons can be seen distinctly, about 30' × 30' or something close. There's no need to zoom in too far or you will lose track of the more distant moons.
- Set TIME FLOW RATE to 1 hour. Press the RUN TIME FORWARD button (▶) and let it run. You will see the four labeled moons move back and forth around Jupiter. Press the STOP TIME button (■) once you have watched several orbits of the moons.

3. Galileo was able to deduce that these four moons moved in nearly circular orbits around Jupiter without the benefit of this time-lapse view or moon labels. Seeing the moons in motion, what evidence can you provide that tells you they do move in a circular motion, even though you cannot see the circle itself?
4. Which moon stays the closest to Jupiter? Which moon moves the farthest from Jupiter?
5. Which moon circles Jupiter with the fastest speed? Which moves the slowest?

- Click the STEP TIME FORWARD button (▶|) until the moon Io is as far to the left of Jupiter as it can get. In this position, Io is at its greatest elongation from Jupiter. Record the date and time under Elongation 1 in the table in question 6 on the answer sheet. Because you are observing from space, the times will be shown as a 24-hour clock.
- Continue clicking the STEP TIME FORWARD button (|◀) until Io returns to the same position. Record the date and time under Elongation 2.

- Repeat these two instructions for Europa, Ganymede, and Callisto.
- Find the amount of time between these two sequential elongations for each moon. This will be the orbital period of the moon.

6. Fill out the table in question 6 on the answer sheet, showing sequential elongations for each moon. It will look like this:

Moon	Elongation 1	Elongation 2	Period
Io	6:00, Jan 27	0:00, Jan 29	1 day, 18 hours
Europa			
Ganymede			
Callisto			

7. Using what you have observed of the moons' motions, state the relation between a moon's distance from Jupiter, its orbital speed, and its orbital period.

Name: _____

Class/Section: _____

Starry Night Student Exercise – Answer Sheet (*continues on back*)
The Moons of Jupiter

1. Table of observations of Jupiter's moons:

Date	Time	Sketch		
1/07/1610	8:00 P.M.	*	* ○	*
1/08/1610	8:00 P.M.			
1/10/1610	8:00 P.M.			
1/11/1610	8:00 P.M.			
1/12/1610	8:00 P.M.			
1/13/1610	8:00 P.M.			
1/14/1610	8:00 P.M.			

2. Looking at your observations, can you propose an explanation for why Galileo may not have immediately noticed all four moons? Use one of your own observations to support your answer.

_____

_____

_____

3. Seeing the moons in motion, what evidence can you provide that tells you they do move in a circular motion, even though you cannot see the circle itself?

_____

_____

4. Which moon stays the closest to Jupiter? Which moon moves the farthest from Jupiter?

_____

5. Which moon circles Jupiter with the fastest speed? Which moves the slowest?

_____

6. Table of elongations and periods:

Moon	Elongation 1	Elongation 2	Period
Io	6:00, Jan 27	0:00, Jan 29	1 day, 18 hours
Europa			
Ganymede			
Callisto			

7. Using what you have observed of the moons' motions, state the relation between a moon's distance from Jupiter, its orbital speed, and its orbital period.

_____

_____

The Rings of Saturn

GOAL

- To investigate the changing appearance of Saturn's rings and how this allows us to make new discoveries in the Saturnian system.

READING

- Section 8.5 – Rings Surround the Giant Planets

The four large Jovian planets in our Solar System share a common set of characteristics, such as large gaseous atmospheres, an abundance of moons, and ring systems.

Saturn's ring system is the largest and the only one visible through Earth-bound telescopes. From our point of view, they appear to be large disks orbiting above the planet's equator. Saturn, like Earth, is tilted relative to the plane of the Solar System. This means that we will see the rings in different inclination angles at different points of the year, depending on the tilt of the planet and Earth's position in our orbit. Approximately twice every Saturn orbit around the Sun, Earth will pass through the plane of Saturn's rings. On these days, from Earth's point of view the rings will have a tilt (inclination) of 0°, so that we will see them edge-on. These **ring plane crossings** are times when astronomers can learn more about the thickness and physical arrangement of the rings. Also, because of the lack of glare from the highly reflective rings, these are ideal times to search for previously unknown moons of Saturn. In this exercise, we will investigate the changing appearance of Saturn's rings as seen from Earth.

SETUP

- Start Starry Night.
- Stop the flow of time with the STOP TIME button (■).
- Set the date in the TIME AND DATE window to read January 1, 2009.
- Open the OPTIONS sidebar.
  - Under LOCAL VIEW, turn off all options.
  - Under SOLAR SYSTEM, turn on PLANETS-MOONS. All other options should be off.
- Click the menu button (≡) next to the SEARCH BAR and select the PLANETS sidebar.
- Type Saturn in the SEARCH BAR.
  - Click the ☑ box to the left of Saturn.
  - Click on the menu button (≡) to the left of the checkbox for Saturn to show a drop-down menu. Click CENTER; this will aim your point of view so that Saturn is in the center of your screen.
- Close the sidebar.
- Set the TIME FLOW RATE to be 1 day. Don't start the flow of time yet.
- At the top right, in the ZOOM box, click the + key to zoom in until the display reads 1' × 1'.

ACTIVITY 1 – FINDING SATURN'S RING PLANE CROSSING

1. On this date, is the inclination of Saturn's rings very great or very small?

- Press the START TIME button (▶) to run time forward and observe the changing inclination of Saturn and its rings.

- You will notice that at one point during 2009, the rings will appear edge-on and nearly vanish. Use the STOP TIME button (■) when you get close to or pass this alignment, then use the STEP TIME FORWARD / BACKWARD buttons (▶|/|◀) as needed to find the exact date.

2. What is the date that Saturn's rings appear edge-on? With the rings in this orientation, how might observations of other nearby objects be affected?

3. What can you say about the thickness of Saturn's rings compared to the planet itself? (Try zooming in using the + or − buttons in the ZOOM window in the upper right to get a closer view.)

ACTIVITY 2 – SATURN'S RINGS AT MAXIMUM INCLINATION

- In the ZOOM window in the upper right, reset your zoom to 1' × 1'.
- Set the TIME FLOW RATE to 5 days.
- Press the START TIME button (▶) to run time forward and observe how our view of Saturn and its rings will change over the next few years.
- When the rings appear to be tilted their maximum amount from edge-on, press the STOP TIME button (■). Use the STEP TIME FORWARD / BACKWARD buttons (▶|/|◀) until you are at the approximate date when the rings are tilted the most.

4. As you watched Saturn, its size appeared to fluctuate from slightly larger to slightly smaller. What could be causing the apparent change in size of Saturn? (Think about the orbits of Earth and Saturn and the fact that this fluctuation repeats every year.)

5. What is the approximate date at which Saturn's rings are tilted their maximum amount from edge on?

6. For this date, describe the width of the rings compared to the diameter of Saturn.

7. Based on the amount of time between no tilt and maximum tilt, predict when you think the next ring plane crossing will occur.

ACTIVITY 3 – FINDING THE NEXT RING PLANE CROSSING

- Press the START TIME button (▶) to continue the flow of time.
- The tilt of the rings will again get smaller until the rings are once more edge-on. Use the STOP TIME button (■) when you get close to or pass this alignment, then use the STEP TIME FORWARD / BACKWARD buttons (▶|/|◀) as needed to find the exact date.

8. What is the next date when Saturn's rings again appear edge-on?

9. What is the interval between the two dates when Saturn's rings appear edge-on? How close is this to the date you predicted in question 7?

Name: _____

Class/Section: _____

Starry Night Student Exercise – Answer Sheet
The Rings of Saturn

1. On this date, is the inclination of Saturn's rings very great or very small?

    _____

    _____

2. What is the date that Saturn's rings appear edge-on? With the rings in this orientation, how might observations of other nearby objects be affected?

    _____

3. What can you say about the thickness of Saturn's rings compared to the planet itself? (Try zooming in using the + or – buttons in the ZOOM window in the upper right to get a closer view.)

    _____

4. As you watched Saturn, its size appeared to fluctuate from slightly larger to slightly smaller. What could be causing the apparent change in size of Saturn? (Think about the orbits of Earth and Saturn and the fact that this fluctuation repeats every year.)

    _____

5. What is the approximate date at which Saturn's rings are tilted their maximum amount from edge-on?

    _____

6. For this date, describe the width of the rings compared to the diameter of Saturn.

    _____

7. Based on the amount of time between no tilt and maximum tilt, predict when you think the next ring plane crossing will occur.

    _____

8. What is the next date when Saturn's rings again appear edge-on?

    _____

9. What is the interval between the two dates when Saturn's rings appear edge-on? How close is this to the date you predicted in question 7?

    _____

    _____

Pluto and Kuiper Belt Objects

GOAL

- To show why Pluto's orbit is like those of many other small bodies in the outer Solar System.

READING

- Section 9.1 – Dwarf Planets May Outnumber Planets

In 2006, there was a lot of publicity about new guidelines from the International Astronomical Union about the use of the word **planet**; as you probably know, Pluto doesn't fit the definition. Pluto and its companion Charon are very small, icy worlds—in their composition, they are more like giant comet nuclei than any other class of Solar System objects.

Astronomers have long suspected that the outer reaches of the Solar System may contain a large number of objects like Pluto in the distant Kuiper Belt. Several of these have been discovered in the past decade or so and are now called **Kuiper Belt objects** (KBOs). In this exercise, we will see that Pluto is similar to the KBOs in its orbit and in its composition.

SETUP

- Start Starry Night.
- Stop the flow of time with the STOP TIME button (■).
- Open the OPTIONS sidebar.
  - Under GUIDES, turn all options off.
  - Under LOCAL VIEW, turn all options off.
  - Under STARS, turn all options off.
  - Under SOLAR SYSTEM, turn on ASTEROIDS, turn on PLANETS-MOONS; turn all other options off.
  - Under CONSTELLATIONS, turn off all options.
  - Under DEEP SPACE, turn off all options.
- Close the OPTIONS sidebar. At the moment, there may be few objects visible, as the program is now showing only those planets that are above the horizon from your viewing location.
- Select OPTIONS / VIEWING LOCATION from the drop-down menus at the top of the Starry Night window.
  - In the subwindow that opens, click on the down arrow (▼) next to the display box for VIEW FROM. Select STATIONARY LOCATION.
  - In the boxes labeled CARTESIAN COORDINATES, enter the following values:
    - X: 0 AU
    - Y: 0 AU
    - Z: 75 AU
  - Click on the GO TO LOCATION button. The view will now shift to a location above the Sun so that you are looking down from above the plane of the Solar System to see the orbits of asteroids and planets. You can hit the space bar during the motion to speed up the process.
- Type "**Sun**" in the SEARCH BAR at upper right.
  - Double-click on the word SUN in the results list. This will shift the angle of your gaze so that it is centered on the Sun.
- Scroll down the results list to find Neptune.
  - Click the ☑ box to the left of Neptune to label it and click the circle to the right to show its orbit.
- Scroll down the results list to find the category DWARF PLANETS, which contains the entry for Pluto.

- ◦ Click the ☑ box to the left of Pluto to label it and click the circle to the right to show its orbit.
- Leave the FIND tab open.
- Zoom in or out using the + and − buttons at the top right of the Starry Night window. Adjust the zoom until the orbit of Pluto fills the display but is still visible all around.

ACTIVITY 1 – PERIHELION AND APHELION OF NEPTUNE AND PLUTO

Pluto can approach closer to the Sun than Neptune does, but its orbit does not cross Neptune's. Explore this by finding the minimum distance (perihelion) for Pluto, and compare it to Neptune's distance at the time Pluto was at perihelion.

- Set the TIME FLOW RATE to 20 days.
- Move your cursor to the word PLUTO in the results display at right. Click on the menu button (≡) to the left of PLUTO, and select SHOW INFO.
- Scroll down to find the POSITION IN SPACE section, which contains information on Pluto's location. If it is not expanded, click the ▶ button next to the category to show the information. Leave the sidebar open.
- Using the TIME FLOW buttons, find the moment when Pluto was last closest to the Sun. (This was back around 1989.) Use the STEP TIME FORWARD (▶|) and STEP TIME BACKWARD (|◀) buttons to stop at the time when Pluto was at perihelion.

1. What was the month and year in which Pluto was at perihelion?
2. What distance was Pluto from the Sun in that month?

- Move your cursor to the word NEPTUNE in the results display at right. Click on the menu button (≡) to the left of NEPTUNE, and select SHOW INFO.
- Scroll down to find the POSITION IN SPACE section, which contains information on Neptune's location. If it is not expanded, click the ▶ button next to the category to show the information.

3. What was Neptune's distance from the Sun during that time?

ACTIVITY 2 – ORBITS IN THE OUTER SOLAR SYSTEM

- In the DWARF PLANETS category along with Pluto are the names of three other dwarf planets that are members of the Kuiper Belt, and so are called Kuiper Belt objects (KBOs). They are Haumea, Makemake, and Eris. (Ceres is located in the main asteroid belt in the inner Solar System; do not use it.)
- Click both the boxes to the left and the circles to the right of each of these three names to show the orbits and positions of these Kuiper Belt objects.
- These orbits are much larger than Pluto's orbit, so we need to zoom out. Using the procedure described earlier for setting the location, move the viewpoint to
    - X: 0 AU
    - Y: 0 AU
    - Z: 150 AU
- Zoom in or out so that the largest orbit is just within the viewing window.
- Set the TIME FLOW RATE to 10 days. Start the flow of time. Let it run while you observe the motion of all the bodies on the display, and stop after several years of the simulation have passed. Stop the flow of time.
- Now change your point of view so that you are aligned with Neptune's orbit. Following the procedure described earlier, change your location to
    - X: 0 AU
    - Y: 150 AU
    - Z: 0 AU

After the view shifts, you should easily see the inclinations of the outer Solar System orbits as seen from along the ecliptic plane.

4. All the classical planets have small values of eccentricity and orbital inclination. Based on the values for the other objects in this exercise, is Pluto more like the KBOs or the classical planets?
5. Given the orbits of Pluto and the KBOs, if you wanted to search the sky for similar objects, would you look in the same part of the sky as you find the classical planets? Why or why not?

Name: _____

Class/Section: _____

Starry Night Student Exercise – Answer Sheet
Pluto and Kuiper Belt Objects

1. What was the month and year in which Pluto was at perihelion?

    _____

2. What distance was Pluto from the Sun in that month?

    _____

3. What was Neptune's distance from the Sun during that time?

    _____

4. Is Pluto more like the KBOs or the classical planets?

    _____

5. Given the orbits of Pluto and the KBOs, if you wanted to search the sky for similar objects, would you look in the same part of the sky as you find the classical planets? Why or why not?

    _____

    _____

Asteroids

GOALS

- To explore the orbits of some objects in the asteroid belt between Mars and Jupiter.
- To examine the orbit of one asteroid that passes near Earth.

READING

- Section 9.3 – Asteroids Are Pieces of the Past

There are many asteroids in the **main asteroid belt** between Mars and Jupiter. In addition, some have orbits that take them near Earth. The latter are called **near-Earth asteroids**.

Figure 9.14 of the textbook shows the orbits for several classes of asteroids. Asteroids have orbits that lie near the plane of the Solar System, but on average their orbits are moderately inclined to that plane. In this exercise, we will animate some of the orbits in that figure and view them from different angles so that we can see the shapes and orientations of their orbits.

SETUP

- Start Starry Night.
- Stop the flow of time with the STOP TIME button (■).
- Open the OPTIONS sidebar.
  - Under GUIDES, all options should be off.
  - Under LOCAL VIEW, turn all options off.
  - Under STARS, turn all options off.
  - Under SOLAR SYSTEM, turn on ASTEROIDS, turn on PLANETS-MOONS; turn all other options off.

- Under CONSTELLATIONS, turn off all options.
- Under DEEP SPACE, turn off all options.
- Change your viewing location so that you are looking down at the Sun and can see the orbits of asteroids and planets:
  - Select OPTIONS / VIEWING LOCATION from the drop-down menus at the top of the Starry Night window. The VIEWING LOCATION subwindow will appear.
  - At the top of the subwindow, click on the down arrow to the right of the dialog box labeled VIEW FROM. Select STATIONARY LOCATION.
  - The box will change so that you can select a fixed location in space. These coordinates are centered on the Sun. Replace the numbers in the boxes labeled CARTESIAN COORDINATES, and enter the following values:
    - X: 0 AU
    - Y: 0 AU
    - Z: 15 AU
- Click on the VIEW FROM SELECTED LOCATION button. The view will now shift to a location above the Sun. You can hit the space bar during the motion to speed up the process.
- In the SEARCH BAR at the upper right, type **"Sun."**
  - Double-click on the word SUN in results. This will shift the angle of your gaze so that it is centered on the Sun.
  - The planets will also be listed below. Turn on the labels for Earth, Mars, and Jupiter by clicking the box to the left of their names.
  - Show the orbit of Earth, Mars, and Jupiter by clicking the circle to the right of their names.
- Zoom in or out using the + and − buttons at the top right of the Starry Night window. Adjust the zoom

until the orbit of Jupiter fills the display but is still visible all around.

ACTIVITY 1 – KEPLER'S LAWS REVISITED

- Set the TIME FLOW RATE to 2 days. Turn on the flow of time with the ▶ button. Let it run for a while, then stop the flow of time with the ■ button.

1. There are two planets that orbit inside Earth's orbit. These are not labeled. Which planets are they?
2. You will note that the orbital periods of the planets are larger if they are farther from the Sun. Which one of Kepler's laws describes this?

ACTIVITY 2 – ORIENTATION OF THE ORBITAL PLANES OF THE ASTEROIDS

- Using the drop-down menus at the top of the display, select OPTIONS / SOLAR SYSTEM / ASTEROIDS. A subwindow labeled ASTEROID OPTIONS will appear. Click and drag the slider bar labeled ASTEROID BRIGHTNESS until it is located to the right. Close the subwindow by clicking the OK button.
- Scroll down in the sidebar to find the category DWARF PLANETS.
- If not already done, expand the list of dwarf planets by clicking on the ▶ button next to the words DWARF PLANETS.
- Display the label and orbit for Ceres as you did with the planets above.
- We will find three other well-known asteroids to display as well.
  - Expand the list of SUPERIOR ASTEROIDS and find Pallas; turn on its label and orbit. Collapse the list when done.
  - Expand the list of MAIN BELT ASTEROIDS and find Vesta; turn on its label and orbit. Collapse the list when done.
  - Expand the list of OTHER ASTEROIDS (this list is below the Dwarf Planets) and find Hygiea; turn on its label and orbit. Collapse the list when done.
- Double-click on the word SUN in the NAME column to center the view on the Sun. (It should already be centered, but this step may be necessary after selecting the orbits above.)
- Turn on the flow of time with the ▶ button. Let it run for a while, then stop the flow of time with the ■ button.

3. Are these orbits all circular, like Earth's orbit? Are any even more elongated than the orbit of Mars? (You may need to click the orbits on and off one at a time to see the orbit shapes more clearly.)

- Turn on the flow of time with the ▶ button.
- While the time is flowing, change the viewing location so that you are looking toward the Sun in the plane of Earth's orbit:
  - Select OPTIONS / VIEWING LOCATION from the drop-down menus at the top of the Starry Night window. The VIEWING LOCATION subwindow will appear.
  - At the top of the subwindow, click on the down arrow to the right of the dialog box labeled VIEW FROM. Select STATIONARY LOCATION.
  - The box will change so that you can select a fixed location in space. These coordinates are centered on the Sun. Replace the numbers in the boxes labeled CARTESIAN COORDINATES, and enter the following values:
    - X: 15 AU
    - Y: 0 AU
    - Z: 0 AU
- Click on the VIEW FROM SELECTED LOCATION button. The view will now shift to a location in the plane of the Solar System.

4. Are the orbits of the asteroids all exactly in the plane of the Solar System or do they have significant tilts?

ACTIVITY 3 – ASTEROIDS THAT PASS NEAR EARTH

- Stop the flow of time with the ■ button.
- Turn off the dots representing the asteroids. Using the drop-down menus at the top of the display, select OPTIONS / SOLAR SYSTEM / ASTEROIDS. A subwindow labeled ASTEROID options will appear. Click and drag the slider bar labeled ASTEROID BRIGHTNESS until it is located to the left. Close the subwindow by clicking the OK button.
- Set the TIME AND DATE so that it reads January 1, 1996.
- Shift back to a viewpoint that looks down on the Solar System.
  - Select OPTIONS / VIEWING LOCATION from the drop-down menus at the top of the Starry Night window. The VIEWING LOCATION subwindow will appear.
  - At the top of the subwindow, click on the down arrow to the right of the dialog box labeled VIEW FROM. Select STATIONARY LOCATION.
  - The box will change so that you can select a fixed location in space. These coordinates are centered on the Sun. Replace the numbers in the boxes labeled CARTESIAN COORDINATES, and enter the following values:
    - X: 0 AU
    - Y: 0 AU
    - Z: 5 AU

- Turn off the orbits and labels for Mars and Jupiter. Keep on the label and orbit for Earth.
    - Turn off the orbits for the asteroids selected in the previous activity.
- If your version of Starry Night has an asteroid named Bacchus in the list, turn on the orbit and name for Bacchus.
- If Bacchus is not present, you can create it using the following procedure:
    - Using the drop-down menus at the top of the Starry Night window, select FILE / NEW / NEW ASTEROID ORBITING SUN. A subwindow labeled ASTEROID: UNTITLED will appear.
    - There will be a dialog box at the top that contains the word UNTITLED. Replace UNTITLED with **Bacchus**.
    - The display will show a series of slider bars under a tab called ORBITAL ELEMENTS. To the right of each slider bar is a dialog box in which you can enter numbers manually. Enter the following numbers:

    > Mean distance (a): 1.07795
    > Eccentricity (e): 0.39494
    > Inclination (i): 9.4347
    > Ascending node: 33.17596
    > Arg[ument] of pericenter (w): 55.202
    > Mean anomaly (L): 191.126
    > Epoch: 2454000.5 (Julian day)

    - Close the subwindow by clicking on the × at the top right corner. You will be asked whether you want to save the new information. Select SAVE.
- Now find the entry under the USER CREATED list called Bacchus. You may have to type "**Sun**" again in the SEARCH BAR and then scroll down in the results.
- Turn on the orbit and label for Bacchus.
- Set the TIME FLOW RATE to 6 hours.

- Start the flow of time with the ▶ button. You will notice that Bacchus and Earth apparently came close together in late March 1996.
- The fact that you are working in this exercise means that Bacchus evidently did not strike Earth in 1996!

5. Come up with a hypothesis about the orbit of Bacchus that would explain why it did not collide with Earth.

- Shift back to a viewpoint that is centered on the plane of the Solar System. Select OPTIONS / VIEWING LOCATION from the drop-down menus at the top of the Starry Night window. The VIEWING LOCATION subwindow will appear.
    - At the top of the subwindow, click on the down arrow to the right of the dialog box labeled VIEW FROM. Select STATIONARY LOCATION.
    - Replace the numbers in the boxes labeled CARTESIAN COORDINATES, and enter the following values:

    > X: −3 AU
    > Y: −3 AU
    > Z: 0 AU

    - Click on the VIEW FROM SELECTED LOCATION button. The view will now shift to a location where you can see the orientation of Earth's and Bacchus's orbits. You can hit the space bar during the motion to speed up the process. Other planets, such as Jupiter, may also be in the view.
- Using the TIME FLOW buttons, run time forward or backward to explore the passage of Bacchus near Earth. The minimum separation was about 15 million km.

6. Was your hypothesis correct? Why did Bacchus miss Earth?
7. Can you speculate as to why the orbits of the asteroids differ from those of the planets?

Name: _____

Class/Section: _____

Starry Night Student Exercise – Answer Sheet
Asteroids

1. There are two planets that orbit inside Earth's orbit. These are not labeled. Which planets are they?

   _____

2. You will note that the orbital periods of the planets are larger if they are farther from the Sun. Which one of Kepler's laws describes this?

   _____

3. Are these orbits all circular, like Earth's orbit? Are any more elongated than the orbit of Mars?

   _____

4. Are the orbits of the asteroids all exactly in the plane of the Solar System or do they have significant tilts?

   _____

5. Come up with a hypothesis about the orbit of Bacchus that would explain why it did not collide with Earth.

   _____

   _____

6. Was your hypothesis correct? Why did Bacchus miss Earth?

   _____

   _____

7. Can you speculate as to why the orbits of the asteroids differ from those of the planets?

   _____

   _____

The Magnitude Scale and Distances

GOAL

- To be able to compare distances to stars using brightness and luminosity information.

READING

- Section 10.1 – The Luminosity of a Star Can Be Found from the Brightness and the Distance (especially the subsection "The Brightness of a Star")

Finding distances to stars is one of the most challenging problems in astronomy. Generally speaking, the farther away a light source is moved, the dimmer it looks. But when comparing stars, the luminosity also needs to be considered, as all stars do not emit an equal amount of energy. Because of the relation between a star's brightness and its luminosity (how much energy it emits per second), we can determine its distance.

Astronomers commonly make use of a scale called *magnitudes* to describe the brightness and luminosity of a star. This scale is set up so that smaller (and even negative) numbers represent brighter or more luminous stars. A star's **apparent magnitude** represents how bright it appears in our night sky. A star's **absolute magnitude** represents how bright a star would look if it were seen from a distance of 10 pc (32.6 light-years). Because the absolute magnitudes of all stars are expressed as if the stars were at that distance, the number is directly related to how luminous a star is. When comparing stars, larger-magnitude numbers mean dimmer or less luminous stars, whereas smaller or more negative numbers mean brighter or more luminous stars. These values can also be used to compare distances. A star with very similar apparent and absolute magnitudes must be close to 10 pc away. For stars that are much farther away, there will be a larger difference in the magnitudes. In this exercise, we will gather magnitude information for the main stars of the constellation Hercules and use this to compare luminosities and distances.

SETUP

- Start Starry Night.
- Stop the flow of time with the STOP TIME button (■).
- Open the OPTIONS sidebar.
  - Under LOCAL VIEW, turn on LOCAL HORIZON. Turn all other options off.
  - Under SOLAR SYSTEM, turn off all options.
  - Under STARS, turn on STARS. Turn all other options off.
  - Under CONSTELLATIONS, turn on BOUNDARIES, LABELS, and STICK FIGURES (ASTRONOMICAL). Turn all other options off.
- Close the OPTIONS sidebar.
- Set the TIME FLOW RATE to 2 minutes.
- Aim your gaze toward the east point of the horizon by typing **E**.
- Run time forward until the constellation Hercules has risen. If it is already up, use the handgrab tool to center Hercules on your screen. Zoom in until the outlined pattern of Hercules fills most of your screen.

ACTIVITY 1 – APPARENT AND ABSOLUTE MAGNITUDES OF STARS IN HERCULES

- Point at one of the bright stars in Hercules connected with the stick figure. The star's name and some other information will pop up on the screen.
- With the cursor pointing at the star, right-click on it. Select the SHOW INFO option from the menu that pops up.
- The star's information will appear in a new window. In this window, scroll down to find the OTHER DATA section. In this section are entries for the star's magnitude (apparent) and absolute magnitude.
- In the table provided on the answer sheet, write down the apparent magnitude and absolute magnitude for the star you have selected.
- Close the information window and proceed to the other stars along the stick figure for Hercules.

1. Record the apparent magnitude and absolute magnitude for each star given in the table on your answer sheet, but do not fill in the BRIGHTNESS and LUMINOSITY RANKING columns yet.

ACTIVITY 2 – USING THE MAGNITUDE SCALE

- On your answer sheet, you should now have apparent and absolute magnitudes for the 18 bright stars in the stick figure of Hercules. You will now use the information to rank the stars in two ways. First, you will use the apparent magnitude numbers to rank the stars from brightest to dimmest. Then you will use the absolute magnitude numbers to rank their true luminosities, again from greatest to least. Remember that the magnitude scale is a reversed scale; refer to the reading at the beginning of this exercise or in the textbook to guide your rankings.

2. In the column labeled BRIGHTNESS RANKING, rank the stars from brightest to dimmest. If two stars have the same brightness, give them the same ranking.
3. In the column labeled LUMINOSITY RANKING, rank the stars from most to least luminous. If two stars have the same luminosity, give them the same ranking.
4. Are the brightest stars necessarily the most luminous? Explain why this may not be the case.

ACTIVITY 3 – COMPARING DISTANCES USING MAGNITUDES

5. Find two stars in the table with the same apparent magnitude. Which star is farther away? Explain how you can tell.
6. Find two stars in the table with the same absolute magnitude. Which one is farther away? Explain how you can tell.
7. Is the brightest star necessarily the closest? Explain why this may not be the case.
8. By comparing the apparent and absolute magnitudes, find which star is closest and which star is farthest. Explain your reasoning.

Name: _____

Class/Section: _____

Starry Night Student Exercise – Answer Sheet (continues on back)
The Magnitude Scale and Distances

1. Apparent and absolute magnitudes of bright stars in Hercules:

Star	Apparent Magnitude	Absolute Magnitude	Brightness Ranking	Luminosity Ranking
Epsilon Herculis				
Eta Herculis				
Gamma Herculis				
Iota Herculis				
Kornephoros				
Lambda Herculis				
Mu Herculis				
Omicron Herculis				
Phi Herculis				
Pi Herculis				
Rasalgethi				
Rho Herculis				
Sarin				
Sigma Herculis				
Tau Herculis				
Theta Herculis				
Xi Herculis				
Zeta Herculis				

2. In the column labeled BRIGHTNESS RANKING, rank the stars from brightest to dimmest. If two stars have the same brightness, give them the same ranking.

3. In the column labeled LUMINOSITY RANKING, rank the stars from most to least luminous. If two stars have the same luminosity, give them the same ranking.

4. Are the brightest stars necessarily the most luminous? Explain why this may not be the case.

_____

_____

5. Find two stars in the table with the same apparent magnitude. Which one is farther away? Explain how you can tell.

_____

_____

6. Find two stars in the table with the same absolute magnitude. Which one is farther away? Explain how you can tell.

_____

_____

7. Is the brightest star necessarily the closest? Explain why this may not be the case.

_____

_____

8. By comparing the apparent and absolute magnitudes, find which star is closest and which star is farthest. Explain your reasoning.

_____

_____

STUDENT EXERCISE | Stars and the H-R Diagram

GOAL

- To gather data on nearby stars and on bright stars and to reproduce the results shown in Figure 10.19 of the text.

READING

- Section 10.4 – The H-R Diagram Is the Key to Understanding Stars

The **Hertzsprung-Russell diagram** (H-R diagram) is a fundamental tool for understanding the structure and evolution of stars. It is a plot of the luminosity of stars against their temperatures. In this exercise, we express the temperature using the Kelvin scale. The luminosity is given with respect to the solar luminosity; a star with luminosity equal to that of 100 Suns, for example, emits 100 times as much energy each second as the Sun.

To find which stars are most common, we look near the Sun and sample all stars in a specified volume of space. When we do so, we find that most stars are found near the **main sequence** on the H-R diagram. Along the main sequence, there is a well-defined **luminosity-temperature relationship**: stars run from hot/high luminosity to cool/low luminosity. Most stars in the local volume of space are cooler and less luminous than the Sun.

In contrast, most of the bright stars in the sky, such as those that make up the familiar constellations, are very luminous. As a result, they can be fairly bright even when located at long distances from the Sun.

SETUP

- Start Starry Night.
- Stop the flow of time with the STOP TIME button (■).
- Open the OPTIONS sidebar.
  - Under GUIDES, turn all options off.
  - Under LOCAL VIEW, turn on LOCAL HORIZON. Turn all other options off.
  - Under SOLAR SYSTEM, turn off all options.
  - Under STARS, turn on STARS; turn all other options off.
  - Under CONSTELLATIONS, turn on BOUNDARIES, LABELS, and STICK FIGURES. All other options should be off.
- Close the OPTIONS sidebar.
- Set the TIME FLOW RATE to 2 minutes.
- Aim your gaze toward the east point of the horizon by typing **E**.
- Run time forward until the constellation Hercules has risen. If it is already up, use the handgrab tool to center Hercules on your screen. Zoom in until Hercules takes up most of the observation window.

ACTIVITY 1 – LUMINOSITIES AND TEMPERATURES OF BRIGHT STARS IN HERCULES

- Point at one of the bright stars in Hercules connected with the stick figure. The star's name and some other information will pop up on the screen.
- With the cursor pointing at the star, right-click on it. Select the SHOW INFO option from the menu that pops up.

- The star's information will appear in a new window. In this window, scroll down to find the OTHER DATA section. In this section are entries for the star's temperature (in Kelvin) and luminosity (in solar units).
- In the table provided on the answer sheet, write down the temperature and the luminosity for the star you selected. The information may not be known for all stars.
- Proceed to the other stars along the stick figure for Hercules.

1. Record the luminosity and temperature for each bright star in Hercules in the table on your answer sheet.

ACTIVITY 2 – LUMINOSITIES AND TEMPERATURES OF NEARBY STARS

- In the OPTIONS sidebar, turn off LOCAL HORIZON under LOCAL VIEW. Close the OPTIONS sidebar. This will allow you to see stars in all directions, even if they are below the horizon.
- Click on the menu button in the upper right next to the SEARCH bar, and select ADVANCED FIND.
- In the ADVANCED FIND window you will be typing the names of several stars into the NAME text box. The option "is exactly" should be selected in the dropdown menu next to NAME.
- On the answer sheet, there is a table with the names of several nearby stars for question 2. For each star in this table type the name into the NAME text box, being sure to type it exactly as it appears in the table. After entering the name, press the Enter

(or Return) key. An entry for the star will appear in the sidebar under SKY OBJECTS.
- Double-click on the star's name in the sidebar. The view will shift to the location of that star. You can press the space bar while the motion is happening to speed up the process.
- Open the SHOW INFO window for the star either by pointing to it with the cursor and right-clicking to bring up the menu or by right-clicking on the star's name in the sidebar. Write down the temperature and the luminosity information in the table provided on the answer sheet.
- Repeat the process for the other nearby stars in the table on the answer sheet.

2. Record the luminosity and temperature for each of the given nearby stars: HIP71683, HIP71681, HIP54035, HIP32349, HIP16537, HIP37279, HIP108870, and HIP439. These are among the nearest stars to the Sun.

ACTIVITY 3 – THE H-R DIAGRAM

3. On your answer sheet, you should now have tables of the luminosity and temperature for (a) several bright stars in the sky and (b) several nearby stars. Plot these data on an H-R diagram, also supplied on the answer sheet. Use different symbols for each group of stars (for example, X's and O's). Include the information for the Sun in the plot for nearby stars.
4. Are most bright stars near the main sequence? What about the nearby stars?
5. Which sample of stars do you think is more representative of "average" stars in the galaxy? Why?

Name: _____

Class/Section: _____

Starry Night Student Exercise – Answer Sheet (continues on back)
Stars and the H-R Diagram

1. Luminosity and temperature of bright stars in Hercules:

Star	Luminosity	Temperature (K)
Epsilon Herculis		
Eta Herculis		
Gamma Herculis		
Iota Herculis		
Kornephoros		
Lambda Herculis		
Mu Herculis		
Omicron Herculis		
Phi Herculis		
Pi Herculis		
Rasalgethi		
Rho Herculis		
Sarin		
Sigma Herculis		
Tau Herculis		
Theta Herculis		
Xi Herculis		
Zeta Herculis		

2. Luminosity and temperature of nearby stars:

Star	Luminosity	Temperature (K)
Sun	1	5770
HIP71683	1.8	5715
HIP71681		
HIP54035		
HIP32349		
HIP16537		
HIP37279		
HIP108870		
HIP439		

3. Plot the data on the H-R diagram of luminous and nearby stars (use a different symbol for each group):

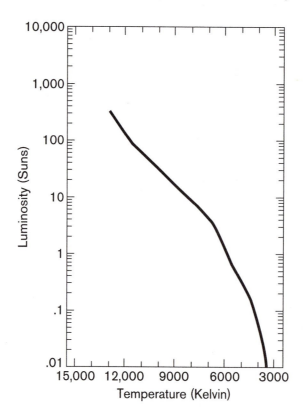

4. Are most bright stars near the main sequence? What about the nearby stars?

_____

_____

5. Which sample of stars do you think is more representative of "average" stars in the galaxy? Why?

_____

_____

Nebulae: The Birth and Death of Stars

GOAL

- To differentiate between the types of nebulae that give birth to stars and those that arise from the death of stars.

READING

- Section 5.2 – The Protostar Becomes a Star
- Section 12.4 – The Low-Mass Star Enters the Last Stages of Its Evolution

Nebulae are among the most beautiful and photogenic objects in space. Basically, a **nebula** is a cloud of gas in space. In spiral galaxies like our Milky Way, clouds of gas are very common, making up a large part of the spiral arms. These clouds are generally dark and diffuse. It is when a cloud is lit up that it makes a more colorful showing. These colorful, showy nebulae come in a variety of types, forms, and colors, each arising from a different process. In this exercise, we will be investigating both **emission nebulae** (also called H II regions) that give birth to stars and **planetary nebulae** that are the result of the end stages of the lives of low-mass stars like our Sun.

SETUP

- Start Starry Night.
- Stop the flow of time with the STOP TIME button (■).
- Open the OPTIONS sidebar.
- Under LOCAL VIEW, turn all options off.

- Under SOLAR SYSTEM, turn on PLANETS-MOONS. Turn all other options off.
- Under STARS, turn on MILKY WAY and STARS. Turn all other options off.
- Under DEEP SPACE, turn on BRIGHT NGC OBJECTS and MESSIER OBJECTS. Turn all other options off.
- Using drop-down menus, select OPTIONS / STARS / MILKY WAY. A subwindow labeled MILKY WAY OPTIONS will open. In this subwindow, click and slide the BRIGHTNESS slider bar all the way to the right. Click OK to close this subwindow.

ACTIVITY 1 – EMISSION NEBULAE

- You will be observing four different emission nebulae: North America Nebula, Eagle Nebula, Omega Nebula, and Lagoon Nebula. For each one, type the full name into the search bar at the upper right and press ENTER. An entry for the nebula will appear in the sidebar.
- Double-click on the nebula's entry in the sidebar. The view will automatically change to center the nebula. You can press the space bar to quickly change your view directly to the nebula.
- Use the + and − buttons as necessary at the bottom left of the screen to zoom in or out to display the nebula's image clearly.
- Repeat the procedure for each of the four nebulae, observing each one and noting any features such as shape, size, color, and so on. You will use your observations to answer the following questions.

1. What color do emission nebulae have in common? Why might they all share this particular color?

2. Do emission nebulae encompass one star or many stars?
3. Describe the locations of these four emission nebulae relative to the band of the Milky Way.

ACTIVITY 2 – PLANETARY NEBULAE

- In the SEARCH BAR at the upper right, type in **M57** and press ENTER. An entry for M57 (Ring Nebula) will appear in the sidebar. Double-click on this entry to center your view on the Ring Nebula, a good example of a typical planetary nebula.
- Use the + and – keys as necessary at the bottom left of the screen to zoom in and out to display the nebula's image clearly.
- Repeat this procedure to search for and observe four other nebulae, typing in their names as follows: **Dumbbell Nebula**; **Blue Snowball**; **Blinking Planetary**; **Eskimo Nebula** (the entry for this will show as NGC 2392). As in the previous activity, observe each one and note any features such as shape, size, color, and so on.

4. Describe the basic shape of planetary nebula.
5. How many stars does each of these planetary nebula contain?

6. Are planetary nebulae larger or smaller than emission nebulae? How can you tell?
7. Compare the colors of the planetary nebulae you have observed with the color of the emission nebulae you observed earlier. What could explain this difference?

- Telescopes in the 1700s and 1800s, when these nebulae were first being identified and named, had a relatively small magnification compared with that of today's telescopes. To simulate this, use the + and – buttons at the bottom left of the screen to set the zoom to $1° \times 1°$.
- In the search bar, type **Jupiter** and double-click on its entry to center your view on this planet. It will be the small disk in the center of your screen. Make sure the zoom stays at $1° \times 1°$, adjusting if necessary.
- Now clear the dialog box and type **M57** and press ENTER. This will center your view on the Ring Nebula. Again, make sure the zoom stays at $1° \times 1°$, adjusting if necessary.

8. After observing both Jupiter and the Ring Nebula at the same low magnification, can you come up with an explanation for why this type of nebula came to be called a "planetary" nebula?

Name: _____

Class/Section: _____

Starry Night Student Exercise – Answer Sheet
Nebulae: The Birth and Death of Stars

1. What color do emission nebulae have in common? Why might they all share this particular color?

    _____

    _____

2. Do emission nebulae encompass one star or many stars?

    _____

3. Describe the locations of these four emission nebulae relative to the Milky Way.

    _____

4. Describe the basic shape of planetary nebula.

    _____

5. How many stars does each of these planetary nebula contain?

    _____

6. Are planetary nebulae larger or smaller than emission nebulae? How can you tell?

    _____

7. Compare the colors of the planetary nebulae you have observed with the color of the emission nebulae you observed earlier. What could explain this difference?

    _____

    _____

8. After observing both Jupiter and the Ring Nebula at the same low magnification, can you come up with an explanation for why this type of nebula came to be called a "planetary" nebula?

    _____

    _____

Pulsars and Supernova Remnants

GOAL

- To investigate the products of supernova explosions.

READING

- Section 12.6 – Binary Stars Sometimes Share Mass, Resulting in Novae and Supernovae
- Section 13.2 – High-Mass Stars Go Out with a Bang

Supernova explosions are the very violent deaths of certain types of stars. Tremendous energy is released during such an event, which will briefly outshine an entire galaxy, making a supernova very visible. There are two basic types of supernovae, depending on the original star. When a white dwarf star (the remnant of a low-mass main-sequence star) is pushed over a certain mass limit, it will detonate in a **Type Ia supernova**, which completely destroys the star. When a high-mass main-sequence star undergoes core collapse, it explodes in a **Type II supernova** that can leave behind the exposed collapsed core. In this case, the exposed core is a superdense compact ball of solid neutrons called a **neutron star**, which is typically the size of a city.

All supernovae leave behind an expanding cloud of debris called a **supernova remnant**. Astronomers generally view the remnants of these explosions, rather than observing the explosion directly. However, even this eventually fades from view as it diffuses itself into the surrounding space, distributing the elements created during the explosion into the galaxy to be incorporated into future generations of stars. Recall that the type II supernova also leaves behind a small, very dense neutron star. The neutron stars created in type II supernovae will often

be spinning at fantastic rates, emitting sweeping beams of radiation from their poles that we see as a **pulsar**. The location of a pulsar tells where a supernova once occurred, even if the remnant debris cloud is no longer visible. In this exercise, we will observe the location and properties of the remnants of several of the more recent of these explosions.

SETUP

- Start Starry Night.
- Stop the flow of time with the STOP TIME button (■).
- Open the OPTIONS sidebar.
  - Under LOCAL VIEW, turn all options off.
  - Under SOLAR SYSTEM, turn all options off.
  - Under STARS, turn on MILKY WAY and PULSARS. Turn all other options off.
  - Under CONSTELLATIONS, turn on BOUNDARIES and LABELS.
  - Under DEEP SPACE, turn on MESSIER OBJECTS. Turn all other options off.
- Close the OPTIONS sidebar.
- Using the drop-down menus, select OPTIONS / STARS / MILKY WAY. A subwindow labeled MILKY WAY OPTIONS will open. In this subwindow, click and slide the BRIGHTNESS slider bar all the way to the right. Click OK to close this subwindow.
- Using the drop-down menus, select OPTIONS / STARS / PULSARS. A subwindow labeled PULSAR OPTIONS will open. In this subwindow, click and slide the NUMBER OF OBJECTS slider bar all the way to the right. Click OK to close this subwindow.

ACTIVITY 1 – PULSARS

- Use the – button at the bottom left of your screen to zoom out to near maximum view. Your screen should look like a circle with the locations of pulsars marked and the band of the Milky Way stretching across. Remember that this represents the sky surrounding you on all sides and that the Milky Way is an encircling ring of clouds.
- Using the handgrab tool, look all around the sky to observe where pulsars lie in the sky relative to the band of the Milky Way.

1. What do you notice about the location of most pulsars relative to the band of the Milky Way? Can you think of a reason to explain this?
2. Which constellation contains the most pulsars? Is this constellation in or near the direction of the center of the Milky Way? Why would this direction contain more pulsars?

- In the SEARCH bar at the top right type in the name **Crab Nebula** and double-click on the resulting entry to center it in your sky. This is the remnant of a supernova, observed nearly 1,000 years ago.
- Use the + button at the bottom left of your screen to zoom in to see the Crab Nebula clearly.
- At the center of the nebula is a green marker representing a pulsar. This is the pulsar at the center of the Crab Nebula. Right-click on the pulsar to bring up a menu of options. Make sure that you have clicked on the pulsar (PSR0534+2200) and not the nebula itself. Select the SHOW INFO option.
- The information for this pulsar will appear in a new window. Look at the information contained in the top section and find "barycentric per." This is the rotational period of the Crab Nebula pulsar in seconds.

3. What is the period of rotation of the Crab Nebula pulsar? How many times does it spin in 1 second?

ACTIVITY 2 – SUPERNOVA REMNANTS

- Open the OPTIONS sidebar.
  - Under DEEP SPACE, turn on BRIGHT NGC OBJECTS and CHANDRA IMAGES.
  - Under OTHER, turn on SUPERNOVA REMNANTS.
- Using the SEARCH bar as you did for the Crab Nebula, find and observe each of the supernova remnants given in the table provided for question 4 on your answer sheet. You will need to continue to adjust the zoom, as the remnants are different apparent sizes. Try to determine if each could be a Type Ia or Type II supernova and record this in the table, along with a brief justification for your decision. Remember that we can distinguish between the two types depending on what they leave behind. In the cases of remnants with pulsars, the images provided by the program may not be exactly lined up with the correct position of the pulsar.

4. Determine the type of supernova remnants listed in the table of the answer sheet and provide a brief justification.
5. Compare the size of the Cygnus Loop (also known as the Veil Nebula) with the other remnants. What does this tell you about its relative age?
6. Do all of these supernova remnants have pulsars in their centers? Why or why not?
7. Do all pulsars have supernova remnants surrounding them? Why or why not?

Name: _____

Class/Section: _____

Starry Night Student Exercise – Answer Sheet (continues on back)
Pulsars and Supernova Remnants

1. What do you notice about the location of most pulsars relative to the band of the Milky Way? Can you think of a reason to explain this?

_____

_____

2. Which constellation contains the most pulsars? Is this constellation in or near the direction of the center of the Milky Way? Why would this direction contain more pulsars?

_____

_____

3. What is the period of rotation of the Crab Nebula pulsar? How many times does it spin in 1 second?

_____

4. Determine the type of supernova remnants listed in the table and provide a brief justification.

Supernova Remnant	Type (Ia or II)	Justification
Crab Nebula		
Tycho		
Cassiopeia A		
Vela Pulsar		
B1509-58		
Cygnus Loop		

5. Compare the size of the Cygnus Loop (also known as the Veil Nebula) with the other remnants. What does this tell you about its relative age?

_____

6. Do all of these supernova remnants have pulsars in their centers? Why or why not?

_____

_____

7. Do all pulsars have supernova remnants surrounding them? Why or why not?

_____

_____

Galaxy Classification

GOAL

- To be able to classify galaxies based on their appearance.

READING

- Section 14.1 – Galaxies Come in Many Sizes and Shapes

In every direction in the sky, when we look beyond the nearby stars of our own Milky Way Galaxy, we see an uncountable number of other galaxies. Edwin Hubble (after whom the Hubble Space Telescope was named) was the first to come up with a classification scheme by observing patterns in a large number of galaxies. Galaxies seem to fall into two basic shapes: spiral and elliptical. *Ellipticals* are classified by how elongated they are, from E0 to E7. *Spirals* are classified by how tight their arms are and how prominent their central bulge is, from Sa to Sc. A spiral galaxy with a barred central bulge are given a separate classification, SBa to SBc. A small number of galaxies have no regular pattern and are called *irregular galaxies*. In this exercise, you will observe many galaxies and learn how to tell the differing shapes apart.

SETUP

- Start Starry Night.
- Stop the flow of time with the STOP TIME button (■).
- Open the OPTIONS sidebar.

- Under LOCAL VIEW, turn all options off.
- Under SOLAR SYSTEM, turn all options off.
- Under STARS, turn on MILKY WAY and STARS. Turn all other options off.
- Under DEEP SPACE, turn on BRIGHT NGC OBJECTS and MESSIER OBJECTS. Turn all other options off.
- Using the drop-down menus, select OPTIONS / STARS / MILKY WAY. A subwindow labeled MILKY WAY OPTIONS will open. In this subwindow, click and slide the BRIGHTNESS slider bar all the way to the right. Click OK to close this subwindow.

ACTIVITY 1 – SPIRAL GALAXIES

- In the search bar at the upper right, type in **M101**. This will bring up several entries under SKY OBJECTS. If no entries appear, open ADVANCED FIND in the menu next to the search bar and type the name there and press FIND. Double-click on M101 (Pinwheel Galaxy). The view will automatically change to center this galaxy. You can press the space bar to quickly change your view directly to the galaxy.
- Use the + and – buttons at the bottom left of the screen to zoom in and out to display the galaxy's image clearly.

1. Describe the basic shape of M101 along with any color differences within the galaxy.
2. What could explain the color difference between the arms and the central core of a spiral galaxy? (Hint: Think about stars of differing colors.)

- In the search bar, erase what is there and type in **M31**. Double-click on M31 (Andromeda Galaxy) to center it.

3. M31 (the Andromeda Galaxy) is another spiral galaxy like M101. What is different about our view of M31 that makes it look so different from M101?

- In the search bar, erase what is there and type in **NGC 4565**. Double-click on its entry to center it.

4. What does this view of a spiral galaxy tell you about the overall shape of the arms and disk?
5. Estimate the thickness of the arms as a fraction of the diameter of the galaxy.

ACTIVITY 2 – ELLIPTICAL GALAXIES

- Now let's look at a couple of elliptical galaxies to compare them to spiral galaxies. In the search bar, erase what is there and type in **M87**. Double-click on M87 to center it.
- Next, in the search bar, erase what is there and type in **M89**. Double-click on M89 to center it.

6. Describe the basic shape of an elliptical galaxy.
7. What color does this type of galaxy appear to have? What does this tell you about the types of stars found in this type of galaxy?

ACTIVITY 3 – IRREGULAR GALAXIES

- In the search bar, erase what is there and type in **LMC**. Double-click on LMC to center it.

8. The Large Magellanic Cloud (LMC) is a satellite irregular galaxy of the Milky Way. What characteristics does the LMC share with either spiral or elliptical galaxies?

ACTIVITY 4 – GALAXY CLASSIFICATION

9. Use the search bar to locate each of the galaxies in the table on your answer sheet. Classify each galaxy using Hubble's tuning fork diagram in Figure 14.2 of the textbook as a guide.

- Close the sidebar.
- Use the – key at top right to zoom out to the maximum view. At this point, you should see the band of our Milky Way stretching across the sky.

10. When we look at our own Milky Way Galaxy, our position inside of it makes it challenging to determine its true shape. How does our view of the Milky Way give evidence of which type of galaxy it is?

Name: _____

Class/Section: _____

Starry Night Student Exercise – Answer Sheet (continues on back)
Galaxy Classification

1. Describe the basic shape of M101 along with any color differences within the galaxy.

    _____

    _____

2. What could explain the color difference between the arms and the central core of a spiral galaxy? (Hint: Think about stars of differing colors.)

    _____

    _____

3. M31 (the Andromeda Galaxy) is another spiral galaxy like M101. What is different about our view of M31 that makes it look so different from M101?

    _____

    _____

4. What does this view of a spiral galaxy tell you about the overall shape of the arms and disk?

    _____

5. Estimate the thickness of the arms as a fraction of the diameter of the galaxy.

    _____

6. Describe the basic shape of an elliptical galaxy.

    _____

7. What color does this type of galaxy appear to have? What does this tell you about the types of stars found in this type of galaxy?

    _____

    _____

8. What characteristics does the LMC share with either spiral or elliptical galaxies?

    _____

9. Classify each galaxy using Hubble's tuning fork diagram. Table of galaxies in the Messier Catalog:

Galaxy	Galaxy Type
LMC	Irr
M87	E0
M101	Sc
M32	
M51	
M59	
M60	
M63	
M64	
M66	
M74	
M82	
M85	
M90	
M104	
M105	
SMC	

10. How does our view of the Milky Way give evidence of which type of galaxy it is?

_____

_____

_____

Quasars and Active Galaxies

GOAL

- To investigate the distribution and luminosities of quasars and active galaxies.

READING

- Section 14.4 – A Supermassive Black Hole Exists at the Heart of Most Galaxies

Among the vast number of galaxies surrounding us in the universe are some that emit incredible amounts of energy, called **active galactic nuclei (AGNs)**. The most luminous of these objects are called **quasars** (the term *quasar* being short for quasi-stellar radio source). These were first detected as radio sources but are now known to correspond to very faint and distant galaxies. The luminosity of quasars is incredible, equaling the combined light of trillions of Sun-like stars, with all of the energy coming from a very compact source in the center of a galaxy.

Other types of AGNs are very similar, but less luminous. It is likely that these are all powered by large and active accretion disks surrounding supermassive black holes in the middle of galaxies that are undergoing some sort of activity like a galactic merger. The interaction drives material into the accretion disk, creating a beacon that can be seen over billions of light years. In this exercise, we will investigate some of the visible properties of quasars and active galaxies.

SETUP

- Start Starry Night.
- Stop the flow of time with the STOP TIME button (■).

- Open the OPTIONS sidebar.
  - Under LOCAL VIEW, turn all options off.
  - Under SOLAR SYSTEM, turn all options off.
  - Under STARS, turn on MILKY WAY and STARS. All other options should be off.
  - Under DEEP SPACE, turn on QUASARS. Turn all other options off.
  - Click the arrow (▶) button next to QUASARS to display other options. Turn off ACTIVE GALAXY and BL LAC OBJECT, and leave QUASARS on.
- Close the OPTIONS sidebar.
- Using the drop-down menus, select OPTIONS / STARS / MILKY WAY. A subwindow labeled MILKY WAY OPTIONS will open. In this subwindow, click and slide the BRIGHTNESS slider bar all the way to the right. Click OK to close this subwindow.
- Using the drop-down menus, select OPTIONS / DEEP SPACE / QUASARS. A subwindow labeled QUASAR OPTIONS will open. In this subwindow, click and slide the NUMBER OF OBJECTS slider bar all the way to the right. Click OK to close this subwindow.

ACTIVITY 1 – QUASARS

- Use the – key at the bottom left of your screen to zoom out to near maximum view. Your screen should look like a circle with the locations of quasars marked with crosses and the band of the Milky Way stretching across.
- Using the handgrab tool, swivel your view around to observe where quasars lie in the sky relative to the Milky Way.

1. Describe the distribution of quasars relative to the Milky Way. Does this imply the quasars are inside or outside of the Milky Way?

- Open the OPTIONS sidebar.
- Under CONSTELLATIONS, turn on BOUNDARIES, LABELS, and STICK FIGURES.
- Close the OPTIONS sidebar.
- Using the handgrab tool, scroll your view around the sky until you find the constellation Ursa Major. Using the + and – keys at the bottom left of your screen, zoom in until as much of the constellation as possible fills your screen.

2. Choose four of the seven quasars in Ursa Major and record their names, redshifts, apparent magnitudes, and absolute magnitudes in the table on your answer sheet. Calculate the average values of both the absolute and apparent magnitudes.

3. Our Sun has an apparent magnitude of about −26.7 in our daytime sky. Compare the absolute magnitudes of the quasars you chose for question 2 (how bright they would look at 10 pc) with how bright the Sun looks in our daytime sky. (Note: You would not want to be 10 pc from a quasar in order to observe this!)

- In a previous exercise ("The Magnitude Scale and Distances"), we used the apparent and absolute magnitudes of stars to estimate their distances. The same thing can be done to estimate the distances to quasars. In the table that follows are a few examples of the magnitudes and distances of objects within our Milky Way:

Name	Apparent Magnitude	Absolute Magnitude	Distance (light-years)
Sirius	0.00	0.55	9
Deneb	1.25	−6.95	1,400
M13	7.0	−8.7	25,000

4. Using the trends seen in the examples given in the table above, compare the apparent and absolute magnitudes of the quasars you chose for question 2 and use this to say something about the distance to these objects. Explain your reasoning using one of your entries as a specific example.

ACTIVITY 2 – ACTIVE GALAXIES

- Open the OPTIONS sidebar.
  - Under DEEP SPACE / QUASARS, turn on ACTIVE GALAXY and turn off QUASARS.
- Close the OPTIONS sidebar.

5. Find the trapezoid shape that represents the bowl of the Big Dipper in Ursa Major. Choose four active galaxies within that part of the constellation and record their names, redshifts, apparent magnitudes, and absolute magnitudes in the table on your answer sheet. Calculate the average values of both the absolute and apparent magnitudes.

6. Compare the absolute magnitudes of these galaxies with those of the quasars you previously recorded. Are active galaxies more or less intrinsically bright than quasars? Explain your reasoning using an example of each.

7. As we did in question 4 above, you can compare the apparent and absolute magnitudes of these AGNs to make a distance estimate. Are AGNs generally closer to or farther from us than quasars? Explain your reasoning using one of your entries as a specific example.

8. Remember that according to Hubble's law, the redshift of distant galaxies tells us approximately how far away they are. Which type of object generally has a greater redshift, quasars or AGNs? What is the relation between redshift and your distance estimates for these two types of objects?

Name: _____

Class/Section: _____

Starry Night Student Exercise – Answer Sheet (continues on back)
Quasars and Active Galaxies

1. Describe the distribution of quasars relative to the Milky Way. Does this imply the quasars are inside or outside of the Milky Way?

_____

_____

_____

2. Choose four of the seven quasars in Ursa Major and record their names, redshifts, apparent magnitudes, and absolute magnitudes in the table. Calculate the average values of both the absolute and apparent magnitudes.

Quasar	Redshift	Apparent Magnitude	Absolute Magnitude
Average of magnitudes			

3. Our Sun has an apparent magnitude of about −26.7 in our daytime sky. Compare the absolute magnitudes of the quasars you chose for question 2 (how bright they would look at 10 pc) with how bright the Sun looks in our daytime sky.

_____

_____

4. Compare the apparent and absolute magnitudes of the quasars you chose for question 2 and use this to say something about the distance to these objects. Explain your reasoning using one of your entries as a specific example.

_____

_____

5. Find the trapezoid shape that represents the bowl of the Big Dipper in Ursa Major. Choose four active galaxies within that part of the constellation and record their names, redshifts, apparent magnitudes, and absolute magnitudes in the table. Calculate the average values of both the absolute and apparent magnitudes.

Active Galaxy	Redshift	Apparent Magnitude	Absolute Magnitude
Average of magnitudes		

6. Compare the absolute magnitudes of these galaxies with those of the quasars you previously recorded. Are active galaxies more or less intrinsically bright than quasars? Explain your reasoning using an example of each.

7. As we did in question 4, you can compare the apparent and absolute magnitudes of these AGNs to make a distance estimate. Are AGNs generally closer to or farther from us than quasars? Explain your reasoning using one of your entries as a specific example.

8. Remember that according to Hubble's law, the redshift of distant galaxies tells us approximately how far away they are. Which type of object generally has a greater redshift, quasars or AGNs? What is the relation between redshift and your distance estimates for these two types of objects?

Views of the Milky Way

GOAL

- To investigate the appearance of the Milky Way in multiple wavelength views.

READING

- Section 4.1 – What Is Light?
- Section 15.1 – Measuring the Milky Way Is a Challenge

Visible light is one type of **electromagnetic radiation**, with a very specific range of wavelengths, representing just a small portion of the entire electromagnetic spectrum. There are many other types of radiation that we cannot see that differ from visible light only in their wavelengths. Other forms of radiation, from longest to shortest wavelength, include radio, microwave, infrared, visible, ultraviolet, X-rays, and gamma rays. Types of radiation that have longer wavelengths than visible light (such as infrared and radio) are less energetic than visible light and are produced by cool objects. Types of radiation that have shorter wavelengths than visible light (such as ultraviolet, X-rays, and gamma rays) are more energetic than visible light and are produced by hotter objects.

Until the mid-20th century, our view of the universe was limited to only what we could see with our eyes—visible light. But today we construct telescopes that are designed to detect all other wavelengths of radiation, and these give us a new view of our surroundings. Starry Night has a compilation of views of the Milky Way taken by many of these telescopes, such as the Chandra X-ray Observatory, the Spitzer infrared telescope, and many others that observe all regions of the electromagnetic (EM) spectrum. In this exercise, we will investigate the Milky Way in a variety of different radiations to learn more about its nature and structure.

SETUP

- Start Starry Night.
- Stop the flow of time with the STOP TIME button (■).
- Open the OPTIONS sidebar.
  - Under GUIDES / ECLIPTIC GUIDES, turn on THE ECLIPTIC to show the plane of our Solar System. Turn all other options off.
  - Under LOCAL VIEW, turn all options off.
  - Under SOLAR SYSTEM, turn all options off.
  - Under STARS, turn on MILKY WAY and STARS. All other options should be off.
  - Under CONSTELLATIONS, turn on BOUNDARIES and LABELS.
  - Under DEEP SPACE, turn all options off.
- Close the OPTIONS sidebar.

ACTIVITY 1 – THE MILKY WAY IN VISIBLE LIGHT

- Using the drop-down menus at the top of the display, select OPTIONS / STARS / MILKY WAY. A subwindow labeled MILKY WAY OPTIONS will appear.
- Click and drag the BRIGHTNESS slider bar all the way to the right.
- The WAVELENGTH should be set at its default of VISIBLE SPECTRUM.
- Click OK to close this subwindow.

- Use the handgrab tool to swivel your view around the sky to observe the appearance of the Milky Way. Remember that as you look around, you are seeing the entire celestial sphere that surrounds you.

1. Describe the appearance of the Milky Way.
2. What does the appearance of the Milky Way tell us about the overall shape of the galaxy?
3. Toward which constellation do you think is the center of the galaxy? Explain why you chose this direction.
4. Is the plane of the Solar System lined up with the plane of the Milky Way or are the two planes at a large angle to each other?

ACTIVITY 2 – THE CENTER OF THE MILKY WAY IN OTHER WAVELENGTHS

- Using the drop-down menus at the top of the display, select OPTIONS / STARS / MILKY WAY. A subwindow labeled MILKY WAY OPTIONS will appear.
- Click on the (▼) button to the right of the WAVELENGTH box. This will bring down a list of other wavelength views of the Milky Way. Select NEAR-INFRARED. This will show you the Milky Way in a longer-wavelength radiation than our eyes can see.
- Click OK to close this subwindow.

5. In the direction of which constellation is infrared (IR) radiation brightest? Does this match with your answer to question 3?

- Using the same instructions as above, observe the center of the Milky Way in all wavelength views.

6. Which types of radiation give us the best view of the center of the Milky Way?
7. Which types of radiation do we not see from the center of the Milky Way?
8. What does this tell you about the conditions at the center of our galaxy?

ACTIVITY 3 – OTHER PARTS OF THE MILKY WAY

- Using the same instructions as described for Activity 2, set the wavelength view to X-RAYS.
- Open the OPTIONS sidebar.
  - Under DEEP SPACE, turn on CHANDRA IMAGES and LABELS.
  - Click on CHANDRA IMAGES to bring up another window labeled CHANDRA IMAGE OPTIONS.
  - Click the checkbox next to LABELS and move the slider bar all the way to the right to MORE LABELS.
  - Click OK to close this window.
- Close the OPTIONS sidebar.

9. Looking around the sky, you'll see some very bright sources of X-rays that are not in the direction of the center of the Milky Way. Find two bright X-ray sources that have labeled names, and identify the constellations they are located in.
10. Observe the same two objects in all other wavelengths. Which other wavelengths do these two objects emit? What does this tell you about the nature of these two objects?

- Open the OPTIONS sidebar.
- Under DEEP SPACE, turn off CHANDRA IMAGES.
- Under OTHER, turn on SUPERNOVA REMNANTS. These are clouds of debris left behind when stars explode violently, and they emit a great deal of high-energy radiation.
  - Click on SUPERNOVA REMNANTS to bring up another window labeled SUPERNOVA REMNANTS OPTIONS.
  - Click the checkbox next to LABELS and move the slider bar all the way to the right to MORE LABELS.
  - Click OK to close this window.
- Close the OPTIONS sidebar.
- Using the same instructions as in Activity 2 above, set the wavelength view to GAMMA RAYS.

11. Describe the distribution of supernova remnants in the sky and the relation between this distribution and the gamma-ray view of the Milky Way.

Name: _____

Class/Section: _____

Starry Night Student Exercise – Answer Sheet (continues on back)
Views of the Milky Way

1. Describe the appearance of the Milky Way.

   _____

   _____

2. What does the appearance of the Milky Way tell us about the overall shape of the galaxy?

   _____

3. Toward which constellation do you think is the center of the galaxy? Explain why you chose this direction.

   _____

   _____

4. Is the plane of the Solar System lined up with the plane of the Milky Way or are the two planes at a large angle to each other?

   _____

5. In the direction of which constellation is infrared (IR) radiation brightest? Does this match with your answer to question 3?

   _____

6. Which types of radiation give us the best view of the center of the Milky Way?

   _____

7. Which types of radiation do we not see from the center of the Milky Way?

   _____

8. What does this tell you about the conditions at the center of our galaxy?

   _____

9. Looking around the sky, you'll see some very bright sources of X-rays that are not in the direction of the center of the Milky Way. Find two bright X-ray sources that have labeled names, and identify the constellations they are located in.

   _____

10. Observe the same two objects in all other wavelengths. Which other wavelengths do these two objects emit? What does this tell you about the nature of these two objects?

_____

_____

11. Describe the distribution of supernova remnants and the relation between this distribution and the gamma-ray view of the Milky Way.

_____

_____

Globular Clusters

GOAL

- To use the distribution of globular clusters in our galaxy to estimate the direction and distance to the center of the Milky Way.

READING

- Section 15.1 – Measuring the Milky Way Is a Challenge

In the early 20th century, there was a debate about the nature of our galaxy. The presiding view was that as the Milky Way obviously surrounded our Solar System on all sides, our Solar System must be either at the center or somewhere very near to it. But it was noticed even in the early 1800s that globular clusters of stars appeared to be concentrated in one particular direction in the sky, rather than evenly distributed in all directions as would be expected if our Solar System were at the center of the galaxy.

If these **globular clusters**—compact clusters of several hundred thousand stars within a span of a few dozen light-years—are not orbiting around the Solar System, then there must be another common center elsewhere in our galaxy. By determining the distances to these clusters and plotting their locations in space, astronomers were able to make a rough estimation of the direction and distance to the center of our galaxy. In this exercise, you are reproducing the type of work that astronomers do when researching a sample of objects. We will use Starry Night to find the distances to a selection of globular clusters in the direction of their highest concentration and make a plot of their distribution to find the location of the galactic center.

SETUP

- Start Starry Night.
- Stop the flow of time with the STOP TIME button (■).
- Open the OPTIONS sidebar.
  - Under GUIDES / GALACTIC GUIDES, turn on the EQUATOR and GRID. Turn all other options off.
  - Under LOCAL VIEW, turn all options off.
  - Under SOLAR SYSTEM, turn all options off.
  - Under STARS, turn on MILKY WAY and STAR CLUSTERS. All other options should be off.
  - Expand STAR CLUSTERS, and turn off OPEN CLUSTER.
  - Under CONSTELLATIONS, turn on BOUNDARIES and LABELS.
  - Under DEEP SPACE, turn all options off.
- Close the OPTIONS sidebar.
- Using the drop-down menus, select OPTIONS / STARS / MILKY WAY. A subwindow labeled MILKY WAY OPTIONS will open. In this subwindow, click and slide the BRIGHTNESS slider bar all the way to the right. Click OK to close this subwindow.

ACTIVITY 1 – THE DISTRIBUTION OF GLOBULAR CLUSTERS

- Using the handgrab tool, click on the screen and drag your view around, looking at the entire celestial sphere, paying attention especially to the band of the Milky Way and the distribution of globular clusters, which are represented as small, light-green circles.

1. Describe the general distribution of globular clusters relative to the visible band of the Milky Way.
2. Toward which constellation(s) does there appear to be the most globular clusters?
3. Does the Milky Way look visibly different in the direction of the constellation you chose? If yes, explain how.

ACTIVITY 2 – THE DISTANCES TO GLOBULAR CLUSTERS

- You will now be typing the names of several globular clusters into the search bar, then finding distance and angle information for each. All of these clusters have been chosen from the dense grouping near the constellation Sagittarius.
- For each globular cluster in the table under question 4 on the answer sheet, type in the name as it appears in the left column of the table (such as NGC 5634).
- The object name will appear in the sidebar. Double-click on the entry to center the cluster.
- The cluster will now be centered and labeled. Point at the symbol for the cluster so that its information pops up on your screen. Right-click on the cluster and choose SHOW INFO from the menu options.
- A new window with information for this cluster will be displayed. Under INFO is an item labeled DISTANCE TO SUN (LY). Write this down in the table.
- Under POSITION IN SKY is an item labeled GALACTIC LATITUDE. This represents how many degrees a cluster is located above or below the

apparent plane of the Milky Way. Write down the first number (degrees) in the table.
- Repeat this process for the other globular clusters in the table on the answer sheet.

4. Complete the table of information for each globular cluster.

ACTIVITY 3 – THE DISTANCE TO THE GALACTIC CENTER

5. On your answer sheet, you should now have a table of distance and angle information for many globular clusters that are grouped in the same direction in our sky. Using this information, plot each cluster on the diagram provided under question 5 on the answer sheet. The diagram is a set of concentric, equally spaced semicircles with the Sun at their center. Each semicircle represents a distance of 5,000 light-years. Galactic latitude is measured in degrees above or below the galactic plane, represented by the horizontal 0° line. It is generally easiest to find the distance first along this line, then follow the semicircles around to the proper angle. The first two globular clusters are plotted for you.
6. Mark the apparent center of your plotted distribution of clusters with a big "X." This is a rough estimate of the position of the center of our galaxy. How many light-years is the Sun from the center of our galaxy according to your estimate?
7. What is the distance to the center of the galaxy as given in the textbook? How does this compare with the value you found?

Name: _____

Class/Section: _____

Starry Night Student Exercise – Answer Sheet (continues on back)
Globular Clusters

1. Describe the general distribution of globular clusters relative to the visible band of the Milky Way.

_____

_____

2. Toward which constellation(s) does there appear to be the most globular clusters?

_____

3. Does the Milky Way look visibly different in the direction of the constellation you chose? If yes, explain how.

_____

_____

4. Complete the table of distances and galactic latitude angles for many globular clusters.

Cluster	Distance (light-years)	Latitude (degrees)
NGC 7099	41,000	30
NGC 5897	24,000	46
NGC 5634		
NGC 6093		
NGC 6121		
NGC 6144		
NGC 6171		
NGC 6235		
NGC 6284		
NGC 6316		
NGC 6333		
NGC 6453		
NGC 6522		
NGC 6544		
NGC 6652		
NGC 6681		
NGC 6723		
NGC 6809		
NGC 6864		

5. On the diagram that follows, plot each of the clusters in the table above according to distance versus galactic latitude. Each semicircle represents a distance of 5,000 light-years. The first two have been plotted for you on the diagram.

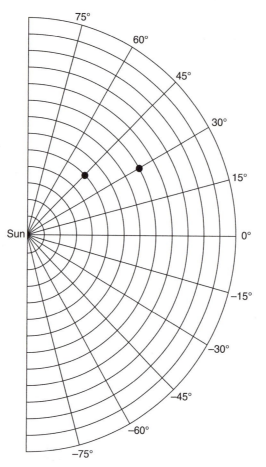

6. Mark the apparent center of your plotted distribution of clusters with a big "X." This is a rough estimate of the position of the center of our galaxy. How many light-years is the Sun from the center of our galaxy according to your estimate?

_____

7. What is the distance to the center of the galaxy as given in the textbook? How does this compare with the value you found?

_____

_____

The Neighborhood of the Sun

GOAL

- To illustrate the concept of high-velocity stars as samples of the local disk and halo of the Milky Way.

READING

- Part of Section 15.2 – Components of the Milky Way Reveal Its Evolution (subsection "Other Halo Components")

Studying the Milky Way in detail requires that we learn about stars at varying distances from the Sun. The presence of dust in the plane of the galaxy is a great complication, as it limits our view along the plane. Astronomers must therefore turn to stars that are near the Sun.

This is less of a disadvantage than it might first appear. Stars near the Sun might be members of the disk: as they orbit the Milky Way, they spend most of their time near the plane of our galaxy. Local stars might also be members of the halo that just happen to be passing near the Sun.

We can find nearby stars in both groups by looking for **high-velocity stars**. These stars are found because their positions on the sky change quickly compared to other stars, although their motions are still slow in human terms. In this exercise, we identify some nearby high-velocity stars and demonstrate their apparent motion.

SETUP

- Start Starry Night.
- Stop the flow of time with the STOP TIME button (■).

- Open the OPTIONS sidebar.
  - Under GUIDES, turn all options off.
  - Under LOCAL VIEW, turn all options off.
  - Under SOLAR SYSTEM, turn all options off.
  - Under STARS, turn on STARS. Turn all other options off.
  - Under CONSTELLATIONS, turn on BOUNDARIES, LABELS, and STICK FIGURES. Turn other options off.
- In the search bar, type in the name **Rigil Kentaurus**.
- Double-click on the entry to shift the view to center on Rigil Kentaurus, a multiple star system also known as Alpha Centauri, the brightest in the constellation Centaurus.
- Close the OPTIONS sidebar.
- Using the + and − buttons at the lower left of the display, set your zoom height to about 10°.
- Set the TIME FLOW RATE to read 999 sidereal days.

ACTIVITY 1 – FINDING NEARBY STARS

- Pick any other reasonably bright star near Rigil Kentaurus. Run the cursor over it until its information pops up. Right-click and select CENTER. The view will shift slightly so that the star you selected is now centered.
- Start the flow of time. The view will change about 2.7 years per time step. This will make the proper motion (the motion of the star relative to other nearby stars in the celestial sphere) of Rigil Kentaurus visible. Run the time forward until about the year 3500 ACE, then stop the flow of time.

1. Consider driving in a car, and imagine looking out the window. The side of the road seems to whiz by at high speeds, but distant mountains will change their apparent position slowly. Would you expect, then, that Rigil Kentaurus is close to the Sun or very distant compared to other stars?

2. What is the distance from the Sun to Rigil Kentaurus? Run the cursor over the star so that its information pops up on the screen. The distance to the star is given in the list of information.

3. Write down the names and distances for three other stars near Rigil Kentaurus in the sky. Did you make the right guess in question 1?

ACTIVITY 2 – THE FASTEST-MOVING STAR

- Double-click on the date to reset it to the current date and time.
- Stop the flow of time with the STOP TIME button (■).
- Type the name **Barnard's star** in the search bar.
  ◦ Double-click on its name to center the gaze on this star. As before, this should show the label of the star. If not, click on the box to the left of the star name.
- Close the sidebar.
- Zoom out using either the mouse wheel or the – button at the bottom left of your display to see the constellation containing this star.

4. What constellation is Barnard's star currently in?

- Zoom in until the value of the height reads about 10°.
- Set the TIME FLOW RATE to 999 sidereal days.
- Start the flow of time. The view will be centered on Barnard's star, so it will seem to be stationary while the background stars move.
- Eventually, Barnard's star will enter a different constellation. You will notice this when the star crosses a line running across the display. Let the time run until the star is well within the new constellation. Stop the flow of time, then zoom out until you can read the constellation names.

5. What constellation will Barnard's star enter about 3,400 years from now?

6. Barnard's star has a higher apparent motion than that of Rigil Kentaurus. Do you expect it is closer or farther away from the Sun than Rigil Kentaurus?

- Zoom in until you can see an image for Barnard's star. You will have to zoom in fairly far because this is a very dim star. Point at the star to display the pop-up information on your screen.

7. What is the distance to Barnard's star?

8. What other factor besides distance determines a star's apparent motion?

Name: _____

Class/Section: _____

Starry Night Student Exercise – Answer Sheet
The Neighborhood of the Sun

1. Do you expect that Rigil Kentaurus is close to the Sun or very distant compared to other stars?

   _____

2. What is the distance from the Sun to Rigil Kentaurus?

   _____

3. Write down the names and distances for three other stars near Rigil Kentaurus in the sky. Did you make the right guess in question 1?

   _____

   _____

   _____

   _____

4. What constellation is Barnard's star currently in?

   _____

5. What constellation will Barnard's star enter about 3,400 years from now?

   _____

6. Barnard's star has a higher apparent motion than that of Rigil Kentaurus. Do you expect it is closer or farther away from the Sun than Rigil Kentaurus?

   _____

7. What is the distance to Barnard's star?

   _____

8. What other factor besides distance determines a star's apparent motion?

   _____

Beyond the Milky Way

GOALS

- To show that even nearby galaxies are located at vastly greater distances than the stars.
- To examine the Milky Way and other galaxies.

READING

- Section 15.1 – Measuring the Milky Way Is a Challenge
- Section 15.5 – The Milky Way Offers Clues about How Galaxies Form
- Section 17.1 – Galaxies Form Groups, Clusters, and Larger Structures

One of the most difficult concepts to appreciate is that objects in the universe are located at tremendously large distances from the Sun. Our brains are not very good at imagining large numbers, though we can deal with them because we have convenient names such as "billion." To explore the size of the universe, we have to move away from the Sun in geometric steps that increase in distance by a multiplicative factor at each step. For example, if we doubled the distance at each step, we step out like this: 1, 2, 4, 8, 16, 32, and so forth.

In this exercise, we will follow a similar pattern, only with much bigger steps. On the way outward from the Sun, we will first show that the stars are vastly farther away than any of the planets in the Solar System. Similarly, all the stars we see at night take up only a tiny part of our galaxy, the Milky Way. The galaxy is, in turn, only one of an immense number of galaxies that stretch out to very great distances. While going through the exercise, remember that the Starry Night program contains in its catalogs only a tiny fraction of all stars and galaxies.

SETUP

- Start Starry Night.
- Stop the flow of time with the STOP TIME button (■).
- Open the OPTIONS sidebar.
  - Under GUIDES, turn all options off.
  - Under LOCAL VIEW, turn all options off.
  - Under SOLAR SYSTEM, turn on PLANETS-MOONS. All other options should be off.
  - Under STARS, turn on STARS. Leave other options off.
  - Under CONSTELLATIONS, turn all options off.
  - Under DEEP SPACE, turn off all options.
- Open the PLANETS sidebar using the menu to the left of the search bar.
  - Click the circles to the right of Earth, Jupiter, and Neptune. This will display the orbits of these planets.
  - Select OPTIONS / VIEWING LOCATION from the drop-down menus at the top of the Starry Night window. Click on the down arrow (▼) next to the display box for VIEW FROM. Select STATIONARY LOCATION.
  - Under CARTESIAN COORDINATES, replace the current numbers with
    X: 0 AU
    Y: 0 AU
    Z: 5 AU

Then hit the VIEW FROM SELECED LOCATION button. The view will now shift so that you are looking down on the Sun and Earth's orbit from a position 5 AU above the Sun.

- Double-click on the entry for the Sun to center on it. You should now see the complete orbit of Earth. Jupiter's orbit might be visible around the edges, and Neptune's orbit is outside your view.
- Close the sidebar.

ACTIVITY 1 – STEPPING TO THE NEAREST STARS

- Using the procedure above, change your location so that you increase your distance from the Sun by a factor of 100. Under CARTESIAN COORDINATES, enter
  X: 0 AU
  Y: 0 AU
  Z: 500 AU

1. The nearest star is located about 300,000 AU away from the Sun. How many times farther away is the nearest star than your current location of 500 AU?
2. From 500 AU, is the orbit of Neptune easily visible? What about the orbit of Earth?

- Open the PLANETS sidebar and turn off the orbits of Earth, Jupiter, and Neptune.
- Close the PLANETS sidebar.
- Now proceed outward by another factor of 100 in distance. Set your location to read
  X: 0 AU
  Y: 0 AU
  Z: 50,000 AU

Then click the VIEW FROM SELECED LOCATION button, but do not press the spacebar. The view will now slowly shift so that you are looking down on the Sun and Earth's orbit from a position 50,000 AU above the Sun.

- While the motion is taking place, look carefully to see if any stars seem to move. You may wish to set the location back to Z = 500 AU and again to Z = 50,000 AU to notice any movement. The apparent movement would be caused by the change in viewing location moving away from the Sun. Only the nearest stars would move.
- If you did see movement of any star, move the cursor to point to the star so that its pop-up information is displayed on your screen.

3. Which stars, if any, moved? What is their distance from the Sun?

ACTIVITY 2 – LEAVING THE MILKY WAY

- Point at the star you selected, and right-click to see the drop-down menu. Click on DESELECT (star

name). The label for this star should disappear. Change your viewing location to
  X: 0 AU
  Y: 0 AU
  Z: 80 ly

This is 100 times farther away than we were before, but we have shifted our units of distance to light-years. During the shift of position, many stars will appear to move.

- Just for fun, set the TIME FLOW RATE to 3,000×. Start the flow of time with ▶, let it run for a while, then stop with ■. This will give you an impression that the stars are scattered through space in three dimensions.
- Now go out another factor of 100 in distance, so that the viewing location is
  X: 0 AU
  Y: 0 AU
  Z: 8,000 ly
- At this point, we are located above the plane of the Milky Way. We can still see dots representing the stars near the Sun in the Starry Night catalog, but notice how compact they are when viewed from this distance. These stars represent a tiny fraction of all the stars in the Milky Way.
- Turn off the individual stars by opening the OPTIONS sidebar and clicking the check box labeled STARS. Also turn off PLANETS-MOONS.
- Under DEEP SPACE turn on 3D GALAXIES.
- In the SEARCH bar, type **Milky Way** and double-click on the entry "Milky Way Centre" that comes up in the results. The gaze will now shift so that you are aimed at the center of the Milky Way galaxy and can see the bright central bulge along with hundreds of other points representing other known distant galaxies. A label for the Milky Way will also appear.
- Click the ∧ button to the left of the VIEWING LOCATION display at the top of the Starry Night window a few times to move yourself even farther away from the Sun until you can see the image of the Milky Way fully. You should be at least 0.2 Mly away at this point.
- Start the flow of time with ▶ and let it run for a while. You will be tracing a big orbit around the Milky Way and can see its shape from different points of view.

4. Is the Milky Way a spiral galaxy or an elliptical galaxy? What features of the Milky Way tell you what type of galaxy it is?

- There are many small dwarf galaxies that orbit the Milky Way as its satellites. As you let time pass and your view of the Milky Way changes, you should be able to identify two satellite galaxies near the Milky

Way, visible as smaller clouds near the Milky Way itself.

- When you find these two small satellite galaxies, stop the flow of time and point at them each with the cursor. The name of the galaxy, along with other information, will pop up on the screen. If you change your view away from the center of the Milky Way you will need to center on it again before continuing to the next exercise.

5. What are the names of two other satellite galaxies?

ACTIVITY 3 – DISTANT GALAXIES

- Let's keep moving out. As you have been doing above, change your viewing location to
    - X: 0 AU
    - Y: 0 AU
    - Z: 2,000,000 ly
- At this distance, you are about as far away as the nearest large galaxy to the Milky Way, the Andromeda Galaxy. Make sure you are still centered on the Sun so that the Milky Way is still in view. Notice how small our own galaxy is beginning to appear.
- Now take yet another step back, by another factor of 100 in distance, so that the viewing location is
    - X: 0 AU
    - Y: 0 AU
    - Z: 200,000,000 ly

You will need to type in the commas or carefully count the zeros. As the distance shifts, the three-dimensional arrangement of distant galaxies is apparent. You may wish to see this by running the flow of time for a while.

6. Do the distant galaxies seem to be uniformly distributed in space or concentrated in groups?
7. What assumptions about the arrangements of galaxies do we make when studying the whole universe? Do you think we are far enough away from the Sun now for these assumptions to be valid?
8. The Hubble constant is 22 km/s/Mly. Using Hubble's law, compute the recession velocity of a galaxy located at our current position (200 Mly from the Sun).

Name: _____

Class/Section: _____

Starry Night Student Exercise – Answer Sheet
Beyond the Milky Way

1. The nearest star is located about 300,000 AU away from the Sun. How many times farther away is the nearest star than your current location of 500 AU?

    _____

2. From 500 AU, is the orbit of Neptune easily visible? What about the orbit of Earth?

    _____

3. Which stars, if any, moved? What is their distance from the Sun?

    _____

4. Is the Milky Way a spiral galaxy or an elliptical galaxy? What features of the Milky Way helped you determine this answer?

    _____

    _____

5. What are the names of two other satellite galaxies?

    _____

6. Do the distant galaxies seem to be uniformly distributed in space or concentrated in groups?

    _____

7. What assumptions about the arrangements of galaxies do we make when studying the whole universe? Do you think we are far enough away from the Sun now for these assumptions to be valid?

    _____

    _____

8. Compute the recession velocity of a galaxy located at our current position (200 Mly from the Sun).

    _____